William W. Morland, Boston Medical Library

The Morbid Effects of the Retention in the Blood

of the elements of the urinary secretion

William W. Morland, Boston Medical Library

The Morbid Effects of the Retention in the Blood
of the elements of the urinary secretion

ISBN/EAN: 9783337390594

Printed in Europe, USA, Canada, Australia, Japan

Cover: Foto ©berggeist007 / pixelio.de

More available books at **www.hansebooks.com**

FISKE FUND PRIZE ESSAY.

THE MORBID EFFECTS

OF THE

RETENTION IN THE BLOOD

OF THE

ELEMENTS OF THE URINARY SECRETION.

BY

WILLIAM WALLACE MORLAND, M. D.,

MEMBER OF THE BOSTON SOCIETY FOR MEDICAL IMPROVEMENT;
ONE OF THE ATTENDING SURGEONS AT THE CENTRAL OFFICE OF THE BOSTON
DISPENSARY, ETC.

BEING THE DISSERTATION TO WHICH THE FISKE FUND PRIZE
WAS AWARDED, JULY 11, 1860.

PHILADELPHIA:
BLANCHARD AND LEA.
1861.

The Trustees of the Fiske Fund, at the annual meeting of the Rhode Island Medical Society, held in Newport, July 11, 1860, announced that the premium of one hundred dollars offered by them on the following subject: "The morbid effects of retention in the blood of the elements of the urinary secretion," had been awarded to the author of the dissertation bearing the motto—

" Prius cognoscere, dein sanare."

And upon breaking the seal of the accompanying packet, they learned that the successful competitor was Wm. W. Morland, M. D., of Boston, Mass.

PHILADELPHIA:
COLLINS, PRINTER, 705 JAYNE STREET.

PUBLISHERS' NOTICE.

DR. CALEB FISKE, who was President of the Rhode Island Medical Society in 1823 and 1824, at his death bequeathed to that Society a fund of two thousand dollars, directing the annual income to be expended in premiums for Essays on subjects selected for competition. The first premium of forty dollars was awarded June 27th, 1836, since which time a large number of valuable dissertations has been laid before the profession through the instrumentality of Dr. Fiske's well-directed munificence. By the judicious management of the Trustees, the Fund has gradually increased, and they are now able to offer two annual prizes of one hundred dollars each.

The Dissertation contained in the present volume received a prize in 1860, and the Trustees have desired that it should be put in a permanent form for consultation and reference, under the belief that it presents a condensed résumé of what is known concerning one of the most interesting pathological questions at present occupying the attention of the profession. It has, therefore, been reprinted from the *American Journal of the Medical Sciences* for April and July, 1861, in which it originally appeared.

PHILADELPHIA, July, 1861.

ON THE EFFECTS OF THE RETENTION

OF THE

URINARY ELEMENTS IN THE BLOOD.

No organs in the human body play a more important part in the economy of life and health than the kidneys—their office is *the depuration of
the blood*. In however slight a degree their function is interfered with,
some untoward effects are produced. These may often be barely noticed,
and easily recovered from ; in many instances, however, although disre-
garded at first, they are sure of their ground, hard to be dislodged, and
too frequently insidious and widely and surely destructive. The more open
and overwhelming attacks of disease, which, by rapidly disabling the kidneys and extensively injuring their tissue, at once and distinctly tell upon
the constitution, reveal in plain characters the close connection between the
vital torrent and its purifying agents.

The subject, as proposed by the Trustees of the Fiske Fund, necessitates,
first, the enumeration of "the elements of the urinary secretion ;" and
secondly, the recital of the effects produced by the undue "retention" of
each of them in the blood.

By the expression "elements of the urinary secretion," as here used, we
understand its constituents in a state of health. These constituents, by a
vital law, are to be eliminated from the blood ; and their retention therein,
beyond a certain time, will certainly cause "morbid effects."

The following enumeration of the urinary elements is taken from one of
the latest and most reliable authorities.[1] The analysis is made up from an
average of the composition of all the urine passed in twenty-four hours.
Average quantity from twenty-four hours, 1400 to 1600 cubic centimetres ;

[1] J. L. W. Thudichum, M. D., Lecturer on Chemistry at the Grosvenor Place
School of Medicine, &c. "A Treatise on the Pathology of the Urine, including a
Complete Guide to its Analysis." London, 1858.

49 to 56 fluidounces. *Average specific gravity,* 1.020. *Mean amount of solids,* 55 to 56 grammes (a gramme is 15.4440 grains, English).

Constituents.

Water	1345 to 1534 grammes.		
Urea	30 to 40	" 463 to 617 grs.	
Uric Acid	0.5	" or 7.5	"
Creatine	0.3	" or 4.5	"
Creatinine	0.45	" or 7.0	"
Sarkine ⎫ Uræmatine ⎬ . . . Uroxanthine ⎭	undetermined.		
Hippuric Acid	0.5	" or 7.5	"
Chlorine	6 to 8	" 92 to 123	"
(or Chloride of Sodium .	10 to 13	" 154 to 200	")
Sulphuric Acid . . .	1.5 to 2.5	" 23 to 38	"
Phosphoric Acid	3.66	" 56	"
Potash and Soda ⎫ Lime and Magnesia ⎭ . . .	undetermined.		
Earthy Phosphates . . .	1.28 grammes,	19	"
Iron	undetermined.		
Ammonia	0.7 grammes,	10	"
Trimethylamine ⎫ Carbonic Acid ⎪ Phenylic Acid ⎬ . . Damaluric Acid ⎭	undetermined.		

"The minor estimates account for 48 out of 55 grammes of solids, the larger estimates for 62 out of 66 grammes of solids."—THUDICHUM, *op. cit.*

From an examination of the above table, in connection with the requisitions of the subject, it will be evident that we have only to indicate the pathological effects arising from the undue retention in the blood, of the following constituents of the urinary secretion : Water ; Urea ; Uric Acid ; Creatine ; Creatinine ; Hippuric Acid ; Chlorine ; Chloride of Sodium ; Sulphuric Acid ; Phosphoric Acid ; Earthy Phosphates, and Ammonia. The other ingredients of the urine, mentioned as being found in "undetermined" proportions, cannot enter into the list, in a practical consideration of the subject.

WATER.—Taking up the urinary constituents in succession, we first examine the results to be observed when that amount of *water* which should be excreted through the agency of the kidneys, is not so evacuated. This portion of our subject may be comprehensively disposed of.

A very variable amount of fluid is evacuated from the bladder at different seasons of the year, and under peculiar and differing circumstances. Thus, in cold weather, the amount of urine is greater, because the cutaneous transpiration is less. Again, when large amounts of liquids are ingested, somewhat corresponding quantities are excreted by the kidneys. The action of abnormally produced sugar occasions diabetes ; certain medicines induce or augment, whilst others restrict, or nearly suspend, the urinary flow. Organic disease, or accidental obstruction, may cause almost complete cessation of urination ; and entire *anuria*, although rare, occurs, from well-known causes.

Whatever, therefore, essentially diminishes, or actually suspends, for a longer or shorter time, the urinary evacuation, causes the retention in the blood of *all* the constituents of the urine, or of a goodly proportion of them. The deleterious effects consequent upon such a retention, will be referable, in the main, to the presence of the solid constituents of the urine, rather than to that of an unusual supply of the watery vehicle. A certain amount—a redundance, even—of water, is absolutely necessary in the circulation, in order to eliminate, wash out, and bear on, as through a sewer, the effete, nitrogenous products, foreign to life, and incompatible with the integrity of the blood. And, besides this necessity for a surplus amount of water, it is rare that enough more than a normal amount is retained in the blood, to be of essential consequence, compared with the effects arising from the presence of the solids of the urinary secretion, prevented from issue by the same cause or causes which retain the watery portion. It is true, however, that "when urea is retained, water is also mostly retained in part, and, by its effusion into the cavities and cellular tissue, causes dropsical disease." (*Thudichum*, op. cit., p. 75.) But the action of the causes just alluded to is rarely or never sufficiently long maintained to be efficient in producing a deteriorated condition of the blood, *referable to excess of water alone*, the kidneys being healthy. Other morbid influences, arising from the presence of the solids of the urinary secretion, and the persistent action of the retaining cause upon the organs themselves, would produce far more rapid and appreciable effects upon the system at large, and upon the blood, than a simple increase of water, only, could do. It is acknowledged, however, that scanty urine—diminished both as to solid and fluid constituents—is indicative of a greater or less degree of anæmia. On the other hand, symptoms of hydruria may be favourable in certain diseased conditions—as where hydræmia and dropsy exist—and its actual establishment, either naturally or by artificial diuresis, may carry off the misplaced water, and restore the balance of the circulation. The profuse flow of watery urine in hysteria is often critical—at all events, *per se*, it indicates no blood-disease. Co-existent anæmia, in such cases, doubtless depends on some other cause than retention of the water of the urine in the blood, or its mere redundance. Often, also, where the quantity of urine excreted is very small, the skin, the bowels, and even the lungs, act *quasi* vicariously, and thus prevent or diminish any ill effects attributable to scantiness of evacuation of the watery portion of the urine. It is well known, also, that the skin will eliminate urea, in cholera, in such quantities that it not only can be detected, but the amount appreciated. (*Thudichum,* op. cit., *et alii.*)

We may now dismiss the watery element from our subject, and proceed at once to the consideration of the undue retention of the solid constituents of the urine in the blood.

UREA.—(Symbol : $\overset{+}{\text{U}}$.—Formula : $C_2H_4N_2O_2$.)[1]

This substance, "the principal product of the metamorphosis in the body of nitrogenized food," and always a constituent of healthy urine, is considered a blood-poison when retained in the circulation. Some observers believe its action to be direct, others that it is indirect—or exerted through the agency of a product of its decomposition. It forms the most considerable portion of the solids of the urinary secretion, and is purely excrementitious matter, the elimination of which by the kidneys is absolutely necessary to health and life. It is true that, in certain exceptional instances, large quantities of it have been ascertained to be present in the blood, for a long time, without compromising life, or even exciting those cerebral symptoms usually observed under such conditions; but it is to be presumed, either that the persons were, to a great extent, insusceptible of the action of urea, or else that the peculiar fermentation supposed to give rise to uræmic poisoning, by producing a noxious substance from the urea, did not take place. It is certainly very possible that some persons may be less impressed by the presence of urea in the blood than others; but, we repeat, such cases must be entirely exceptional. With regard to the constancy of decomposition of urea when retained in the blood, and the consequent formation of another and a toxic substance, we have, as yet, too few facts to enable us to determine. If ever proved to be the rule, however, the intervention of certain unknown agencies might, in isolated instances, prevent its execution ; and thus account for an apparent, or at least a temporary, immunity from morbid consequences.

After extirpation of the kidneys in animals, and in Bright's disease and some other affections, urea is found pervading many of the fluids of the body—as, the dropsical effusion, the blood, the perspiratory secretion, the vitreous and aqueous humours, and the liquor amnii.[2] Dr. Thudichum, who

---

[1] Bird, Thudichum, *et al.*

Chemical Composition of Urea.

	THUDICHUM.	G. BIRD.
2C	20.000	$C_2,N_2,H_4,O_2 = 60.$
4H	6.666	
2N	46.667	
2O	26.667	
	100.000	

[2] The presence of urea in the fluids of the body was first announced by Dr. Christison in the Edinburgh Medical and Surgical Journal, October, 1829.

refers to its detection in the latter fluid by Wöhler, considers its presence there as exceptional; and is inclined, moreover, to throw doubt upon many of the reported instances of its occurrence in other fluids—the reports being in several instances merely assertions by the authors, and not ratified by proof, or else erroneously or partially quoted. Urea has also been declared to have appeared in the milk,[1] in the serum from blisters, and in the alvine evacuations of patients with diseased kidneys.[2]   Dr. Rees states that he has "found most unequivocal evidence of its presence in peritoneal, pericardial, and pleural effusions, and also in the fluid of the arachnoid."[3]

With regard to the question, already alluded to, whether urea retained in the blood is directly or indirectly deleterious—that is, whether it acts *per se* as a poison, or becomes such by a process of decomposition, in which case the carbonate of ammonia is believed to be the injurious agent—there has, of late, been much discussion. The latter view has its zealous advocates, and their theory seems to be somewhat gaining ground. As we have already intimated, there are significant facts adverse to the conclusion that urea alone, as such, is a blood-poison. Dr. Bright remarked that urea may long exist in the blood, in renal disease, and yet no cerebral symptoms arise until the very last of life. He mentions one case which lasted from four to five years. Dr. Rees gives even stronger testimony. He found, in a patient who had no uræmic symptoms whatever, but who retained his cerebral functions to the last moment of his life, the blood more highly charged with urea than he had ever known it in Bright's disease. Dr. Johnson, of London, in his justly celebrated work on diseases of the kidneys, affirms that no actual proof exists that urea is the poisonous agent, or, at least, that it is the only one. If admitted to have a poisonous influence, he holds that some peculiar, unknown condition of the blood must exist, to favour its toxic action. Frerichs is the author of the theory that carbonate of ammonia, resulting from a decomposition which the urea undergoes in the blood, is the poisonous agent. Its presence in the blood he indubitably ascertained, and by injecting it into the bloodvessels of dogs, he produced convulsions. Dr. Hammond, U. S. A., has made some interesting experiments, with the intention of testing this matter. He injected urea, vesical mucus, sulphate of soda, nitrate of potash, and carbonate of ammonia into the blood of dogs; in some instances removing the kidneys previous to injecting the substances. He did not detect ammonia in the breath of any of the animals operated on with *urea* by injection. He inclines to pronounce its presence in Frerichs's cases purely accidental. The animals from whom the kidneys were removed all died, after strong convulsions; and Dr. H. infers an analogy between animals deprived of the kidneys and patients affected with

---

[1] Rees, Diseases of the Kidney, London, 1850, p. 46.   Albuminuria existed in the patient.

[2] Dr. Golding Bird—after the action of elaterium.—*Urinary Deposits.* Loc. cit.

Bright's disease. Such analogy, it is true, may be predicated; but many attendant circumstances attaching to the cases of persons with diseased kidneys do not, of course, affect animals *without* kidneys, and consequently the analogy is not perfect, and hardly as safe to reason from as even the proverbially insecure foundation derived from analogical reasoning generally. Dr. Hammond does not find that urea or carbonate of ammonia, injected into the bloodvessels of sound animals, causes death; if they have suffered extirpation of the kidneys, such injection proves fatal. He does not discover from his experiments that urea, introduced directly into the circulation, becomes converted into carbonate of ammonia.[1] The experiments, at all events, go to prove the deleterious agency of urea, or of the product of its decomposition, when not promptly excreted from the blood, whether it be due, as in the case of the animals experimented upon, to loss of the kidneys, or, as in certain conditions in the human subject, to its retention in the blood by diseased, perverted, or obstructed action of those organs. While the question as to the exact material acting poisonously is still in abeyance, the *facts* relative to urea retained in the blood as productive of various "morbid effects," are indisputable, and as such we shall now proceed to examine them.[2]

GENERAL PHENOMENA REFERABLE TO THE PRESENCE OF UREA IN THE BLOOD.—From the fact that Bright's disease is the affection in which urea is most frequently retained in the blood, it will be all the more necessary not to refer any of its concomitant phases to the action of urea solely; although it is very plausible, and some of the best medical observers of the present day are beginning to teach, that many of the so-called *sequelæ* of Bright's disease may legitimately be referred to the presence of urea in the circulation. Thus, Dr. Watson remarks the extreme readiness of various organs of the body to become inflamed during an attack of Bright's disease. Especially is this found to be true, as all observers will testify, in reference to the serous and mucous membranes. Dr. Watson himself calls attention to this fact, and cites Drs. Bright, Christison, and Gregory to the same effect. He mentions, also, that M. Solon does not, in his volume on albuminuria, consider this tendency especially prominent in France.

Bronchial, pleural, pericardial, peritoneal, gastric, and intestinal inflammations are well-known and common *sequelæ* of Bright's disease, and occur, as to frequency, very nearly in the order above named. Now, it is very plausible to suppose that the abnormal condition of the blood, caused by the presence of urea, may be productive of many of these manifestations. Dr. Watson, while suggesting this, speaks particularly of disorder of the stomach and bowels, which so often follows or is concomitant of

---

[1] North American Medico-Chirurgical Review, March, 1858.
[2] See Appendix, note A.

Bright's disease, and considers it may be explained by the action of "the poisonous material retained in the blood, and seeking a vent through supplementary channels of excretion."[1] He then refers, as corroborative proof, to the *post-mortem* appearances observed in these cases; "most commonly evident traces of disease are met with in various organs" besides the kidneys. This distinguished observer adds, that these manifestations "prevail with irregular frequency in different places. They are probably determined, in some measure, by local and peculiar agencies. Thus, vomiting and diarrhœa have been more familiar to the Edinburgh observers, than in London to Dr. Bright, or in Paris to M. Solon; while the headaches and coma so often witnessed by the British physicians have been comparatively uncommon in France."[2]

Although, in abnormal retention of urea in the blood, the vital tissues and fluids are all more or less affected, and, by a concomitant disturbance of the watery as well as of the solid portions, dropsical effusion may arise, yet the spinal cord and brain are the organs chiefly affected whenever urea becomes a blood-poison; and we have seen that the instances where it is not thus morbidly efficient, when retained in the blood, are exceptional. Generally speaking, when the amount of urea thus traversing the system is considerable, its effects are decided and rapidly disastrous. The affection is, properly, a cachectic condition; in other words, the system is, throughout, evidently depressed by a poison.[3] Were it not, moreover, that the disease of the cerebro-spinal system, consequent on uræmia, is usually so severe, persistent, and fatal, the blood would become very seriously altered in its constituent parts, and finally devitalized, by the retention of urea; and, in addition, the whole play of the vascular system would be disturbed. The large quantities of albumen often eliminated during the retention of unusual amounts of urea in the blood cannot be said to be referable to the presence of urea alone. Albuminuria and uræmia may coexist, but the union is not a necessary one; neither is directly causative of the other. This fact we find pertinently referred to by Prof. G. S. Bedford, in his short but instructive chapter on uræmia.[4]

---

[1] M. Claude Bernard (Leçons sur les Propriétés Physiologiques et les Alterations Pathologiques des Liquides de l'Organisme. Paris : J. B. Baillière, 1859) remarks that in renal disease, when the urine is suppressed, intestinal disorder supervenes. The bowels have taken up, so far as they can, the elimination of urea. The rule is, that there is an elective affinity manifested by certain glands in the elimination of certain products from the blood. When anything interferes with their action, others fulfil their office to the best of their ability.

[2] Lectures on the Principles and Practice of Physic, fourth edition, London, 1857, vol. ii. p. 682.

[3] Notwithstanding that many of the cerebro-spinal phenomena observed are those of irritation or excitation.

[4] Clinical Lectures on the Diseases of Women and Children.

If we were to investigate thoroughly the disturbed chemistry and proportions of the blood which might properly be imputed to retention therein of the solid constituents of the urine, the limits of an essay, such as the requirements of the question now proposed seem to demand, would be soon attained, and, by the addition of the practical details, largely exceeded.

We have interpreted the terms of the subject as indicating a desire for an exposition of the phenomena of disease believed to be legitimately referable to the abnormal retention of the solid urinary constituents; and whilst endeavouring to present these, most of the results of the disturbed proportions and composition of the blood will, in fact, be made evident. And, in concluding these general remarks, it is well to say that accomplished observers, some years since, have been inclined to ascribe many manifestations of disease of slight intensity, and previously obscure and very imperfectly understood, to the presence of abnormal quantities of urea in the blood.[1]

I. Cerebro-Spinal Phenomena attributable to Urea retained in the Blood.

—(A.) A drowsy condition is often the first distinctly declared manifestation of toxæmia by the presence of urea or of the product of its decomposition in the blood. There are sometimes premonitory symptoms of less clear significance, but not always. Some writers, however, class them among the recognized *prodromata*.[2] This drowsy state ordinarily deepens into stupor and true coma, if it be impossible to relieve the blood of the offending element. The re-establishment of free diuresis, or a tendency to recovery from renal disease, may effect so desirable a result; but, unfortunately, the tide too frequently sets in the opposite direction. Coma may prove the final phase of the affection, and be simple or unaccompanied by disordered motility; or it may be combined with convulsions, and life may be terminated even more rapidly than if the complication had not existed.

(B.) Convulsions of an epileptic form may be the sole manifestation; there being no sopor, and consciousness being intact. The state is, in every respect, fully as unpromising as either of the two just indicated.

Among the first symptoms of uræmia may be mentioned—œdema in

---

[1] Urea is always present in healthy blood, but in very small proportion. It may sometimes even not be readily recognized. Morbid effects, consequently, depend upon its presence in large quantities, and upon its *accumulation*, in the blood. The abnormal increase, therefore, even if small, will exert an influence. For valuable chemical and physiological information on this point, see the works of Simon, Thudichum, and Carpenter.

[2] These will be specified hereafter; they are of a character less arrestive of attention than the others; we will only mention—confusion of ideas, failure of memory, unusual sluggishness, general *malaise*.

various parts of the body;[1] lowness of spirits, amounting at times to me-lancholia; restlessness; dizziness; headache; fretfulness; partial anæsthesia and delirium. Nausea, retching, vomiting, and rigors, are likewise noticed. There are often, also, impaired vision, amblyopia, muscæ volitantes, and amaurosis. These latter symptoms, as indeed all the others, are especially mentioned by Dr. Braun, of Vienna, in his late work on Midwifery; a chapter from which, devoted to Uræmic Eclampsia, has been ably trans-lated by Dr. Duncan, of Edinburgh, and has furnished us with a great deal of new and valuable information. While Dr. Braun's views with regard to the subject of uræmic eclampsia will, doubtless, not be at present re-ceived as a whole by the profession, and are, indeed, questioned in many points by his translator, yet his extensive research, accurate observation, admirable description, and ingenious reasoning, render the chapter to which we refer at once entertaining, instructive, and full of practical suggestions. We acknowledge our great indebtedness to the author and his translator.

When stupor and coma finally supervene, a greater or less degree of apoplectic stertor accompanies the respiration. It is uniformly noticed that this stertor has a peculiarly *high tone*—a sort of shrillness, distinguishing it from ordinary apoplectic snoring. Reference was first made to this fact by Addison (*Guy's Hospital Reports*, 1839, No. VI.), and is repeated by Reynolds (*Diagnosis of Diseases of the Brain*, &c., London, 1855) and by Rees (*On Diseases of the Kidney*, 1850).[2]

[1] This peculiar feature of the affection deserves especial notice. Its seats are chiefly the upper part of the body, the face, and the extremities—both upper and lower. The *labia majora* not infrequently exhibit it. Change of the patient's position often causes its disappearance, temporarily; and it frequently becomes less marked, or even vanishes, towards the end of pregnancy, even while the albumen of the urine and the structural disease of the kidneys is increasing. (Braun.) "The skin of the non-œdematous parts of the body appears very dry, and as white as chalk (chlorotic, hydræmic, leukæmic), and has a low tempera-ture. Only those œdemata of pregnant women which exist contemporaneously with albumen, fibrin cylinders, and fatty degenerated scales of Bellini's epithelium in the urine, have a connection with uræmic eclampsia. The œdema of the lower extremities, ascites, and hydramnios, which are not complicated with albuminous urine containing fibrin cylinders, are not followed by uræmic eclampsia in preg-nancy and labour. The affection of the kidneys with disease cannot certainly be inferred from the appearance of dropsy, as distinct causes may, at the same time, or one after the other, produce dropsies." (Braun, *Uræmic Eclampsia;* Duncan, p. 17.)

[2] Dr. Reynolds, commenting upon this characteristic, writes thus: "The stertor exhibits a peculiarity first noticed by Dr. Addison. It is not of low, guttural tone, but of much higher pitch, and appears to be caused by the mouth rather than the throat, either by some position of the tongue against the roof of the mouth or teeth, or by some movement of the arches of the palate, not like that causing ordinary stertor, from which (although its mechanism is obscure) it presents the most obvious difference. (In several obscure cases—*i. e.*, obscure from the fact of the

The action of the poisonous agent in urœmia is believed by the best authorities (Tyler Smith, Braun, Reynolds, Churchill, *et al.*) to be first directed to the spinal marrow; and hence the sensitive impressions which make themselves morbidly apparent, as dizziness, headache, and subsequently convulsive movements. We are not to inquire into the *modus operandi* or etiology of urœmia, or of the other diseased conditions supervening in the human subject, upon retention of the elements of the urinary secretion in the blood; but—as we understand the question—to state the "morbid effects" only; consequently, we shall not occupy time and space by setting forth the received views and theories as to the direct or indirect modes of transmission of the deleterious influences, but will endeavour to state succinctly the disordered vital phenomena observed, and the pathological appearances, if any, which are noted *post mortem.*

Urœmic symptoms may, of course, arise in both sexes from renal disease, or from mechanical obstruction to the excretion of the urine, as in hydronephrosis, retroversion of the uterus, urethral stricture, and closure of the ureters; which latter, if dependent on an unrelievable cause, must soon prove fatal. The occurrence of what has been termed "urœmic eclampsia" has been witnessed in non-pregnant females, and in males, and so cannot be considered as invariably belonging to the parturient state when urinœmia exists. The portion of Dr. Braun's work already cited, is devoted to the exposition of his belief that the convulsions observed during pregnancy are almost exclusively dependent on urinœmia.

After the appearance of the premonitory and of the earlier symptoms of urœmia, the progress of the mischief will, of course, be variable in different patients, and also according to the amount and *cause* of the retention of urea. Thus, it would seem natural that a large amount of urea being somewhat *suddenly* thrown into the circulation, and kept there by the continuance of the cause, should prove rapidly disastrous, and be accompanied with marked and violent phenomena. When gradually introduced, as in the

patient's not having come under notice until cerebral symptoms had appeared and consciousness was so far lost that no commemorative history could be obtained, and in which no œdema of the ankles was perceptible—this peculiarity of the respiratory stertor has at once awakened my suspicions ; has led to an examination of the urine and the breath, and to the discovery in the former of albumen and fibrinous casts, and in the latter of an undue quantity of ammonia.") (*Op. cit.*, p. 110.) The latter fact is significant, in view of the doctrine of Frerichs as to the agent which proves poisonous in urœmia ; and the experiments and observations of many others go far to confirm the opinion.

Dr. Reynolds adds : "The peculiar muscular condition causing this stertor, I am disposed to consider as the result of spasm rather than paralysis, and the spasmodic contraction may be either of sensori-motor, simply reflex, or tonic origin, forming only one of many phenomena which indicate excessive or perverted conditions of those groups of motor action. This hypothesis is, of course, as unimportant as the fact of the difference is valuable."

slower advances of renal disease, or by the action of a progressive obstruction, the system may become somewhat accustomed to the presence of the deleterious agent. May not this be, in some degree, the explanation of the innocuousness of those very considerable amounts of urea the presence of which in the blood, and for a prolonged period, was ascertained by such accurate observers as Bright, Christison, Frerichs, and Rees; in conjunction, as we have previously intimated, with a possible greater power of resistance to the urea-poison in some constitutions than in others ?[1]   If this explanation be not in any degree admitted, the only alternative seems to be to accept the theory of Frerichs, that carbonate of ammonia is the toxic agent.   In support of this view, we have the experiments upon animals, already referred to, in part, where extirpation of the kidneys was practised—as by Prevost and Dumas, Segalas, Tiedemann, Gmelin, Mitscherlich, Claude Bernard, Barreswil, Stannius, and Frerichs, all cited by Dr. Bedford to prove this point (op. cit.); and the test by injection tried by Bichat, Courten, Gaspard, Vauquelin, Segalas, Stannius, Frerichs; both methods without inducing convulsions. (Idem.)   Dr. Bedford also mentions the significant fact that Vauquelin and Segalas proposed to give urea as a diuretic, so little did they consider it a poison! It is, under the present aspect of the subject, as well not to try the experiment.

Orfila—to come to direct experiments—caused fatal convulsions in an animal by the administration of carbonate of ammonia ; and Bernard and Barreswil found carbonate of ammonia in the stomach and intestines of animals after extirpation of the kidneys.

Dr. Rees's idea that a peculiar "tenuity" of the blood may be requisite, in order to have full toxæmic action, when urea is retained, is certainly plausible; for we may at least suppose that the poisonous matter will be more readily and abundantly distributed through the circulating medium, and will consequently more thoroughly pervade and act upon the system.[2]   And here we cannot refrain from adducing the exceedingly acute and ingenious remarks made upon this point by Prof. Simpson, of Edinburgh (Obstetric Works, vol. i. p. 371, American edition), in the article containing his statements in reference to puerperal convulsions, which latter, as we have already mentioned, recent observers have distinctly referred, in a large majority of cases, to toxæmia by the retention of urea, or of the product of its decomposition, in the circulation.   In this particular connection, however, the patients were children—so that here we have remarkable instances of direct

---

[1] Frerichs states that the presence of the as yet unknown ferment in the blood is necessary, in order to the production of toxæmic symptoms by generation of the carbonate of ammonia.   He thus explains the toleration of so much urea in certain cases.

[2] Dr. Todd ("Lumleian Lectures on Delirium and Coma," Med. Gaz., 1850) also favours this idea.   He believes the poisonous action of urea is facilitated by impoverished blood.

uræmic poisoning in connection with albuminuria—convulsions being the prominent symptom. The account of the first case we transcribe entire, together with a foot-note of much interest.

"A few weeks ago, I saw an instance in which convulsions in a child after birth were connected with the presence of albuminuria in its urine; or connected, as it should be, perhaps, more correctly stated, with that condition of the blood-poisoning or uræmia which is the result of albuminuria—whether that condition consists in a morbid accumulation of urea, or is produced, as Frerichs supposes, by the presence of carbonate of ammonia in the blood, produced by decomposition of the urea, or is, as is more probable, the effect of some other morbific agent in the circulating system, capable, like strychnia, of increasing the centric irritability or polarity of the spinal system to such an excessive degree that, under this super-excitability, comparatively slight eccentric causes of irritation in the stomach, intestines, uterus, bladder, &c. &c., readily induce convulsive attacks of a general form, like those of puerperal eclampsia." (*Loc. cit.*)

In the foot-note appended to the above passage, and in reference particularly to the theory of Frerichs, Professor Simpson makes the following important and interesting suggestions:—

"If the blood-poison, which in albuminuria produces convulsions and coma, be, as Frerichs believes, carbonate of ammonia, resulting from decomposition of urea, can we account for the power of chloroform in restraining and arresting, as it does, puerperal convulsions, upon the ground of its preventing this decomposition? The inhalation of chloroform produces, as various chemists have shown, a temporary diabetes; sugar appears in the urine, and hence, probably, also in the blood. The addition of a little sugar to urine *out* of the body, prevents, for a time, the common decomposition of its urea into carbonate of ammonia."

After mentioning the death of another child from convulsions supervening on the third day after birth—the mother having had puerperal convulsions and recovered—Dr. Simpson states that Dr. Weir, of Edinburgh, and himself found the urine of the child, like that of the mother, highly albuminous. He also says he is unaware of any reported observation of the coexistence of albuminuria and infantile convulsions; and then hints at the possibility that the albuminuria may be common as a pathological condition in certain forms of the convulsions of infants—as in trismus nascentium. Other infantile diseases, he thinks, may be powerfully influenced by albuminuria— as, for instance, sclerema, the "endurcissement ou l'œdeme du tissu cellulaire" of French writers. Dr. Simpson had only seen two cases of this in Edinburgh, but was led, at the time of observing them, to believe and "to suggest that the skin-bound disease itself, or at least some forms of sclerema, may be a variety or effect of Bright's disease in early infancy; the effusion into the cellular tissue, which constitutes the marked feature of the affection, being so far analogous to the anasarca occurring with albuminous nephritis."

In reference to the use of chloroform, and the explanation which Dr.

Simpson attaches to its mode of action in overcoming puerperal or uræmic convulsions, may we not ask whether the subduing power of the anæsthetic agent, acting as it does upon the cerebro-spinal system, directly, is not sufficient, of itself, to explain the control of the convulsive manifestations, without a resort to the exceedingly ingenious suggestion of Professor Simpson as to the chemical explanation of the result?

We may here remark that, in one case, Dr. Duncan of Edinburgh found that the inhalation of chloroform aggravated the stertor and lividity of countenance observed in a case of puerperal convulsions ("uræmic eclampsia" of Braun and others). We observe that the chloroform was administered in "small quantity"—perhaps Dr. Simpson might say the amount was not sufficiently large.

In a valuable note to a portion of the chapter of Dr. Braun's work which he has translated, Dr. Duncan has virtually enunciated the same opinions as Dr. Simpson's, previously cited—both in reference to increase of the nervous irritability acting on various organs, and to the analogy of action to be predicated from the experiments instituted by zealous students of these phenomena upon animals. We append his comprehensive and apposite remarks :—

"In uræmia, the most important point is the circulation of a morbid fluid in the nervous system, which probably does not act as a direct excitant of the convulsive motions, so much as it increases the irritability of the nerves, and the consequent liability to convulsions from exciting causes, which, under other circumstances, would produce no noticeable disturbance. Ingenious experiments have, as is well known, been performed on frogs, which seem to demonstrate an analogous condition to exist under poisoning by strychnia, at least when moderate quantities of the poison are administered." (*Loc. cit.*, pp. 59, 60.)

When the peculiar conditions to which the retention of urea in the blood is due, can be relieved and removed, we may witness rallying, and final recovery, even from very unpromising states. Persistence of the cause, however, by maintaining the presence and increasing the amount of the poison, soon induces the gravest accidents, and must terminate fatally, sooner or later, according to the violence of the attack, and the power of resistance manifested by the patient. It will serve at once the purpose of illustrating this fact, and of furnishing a synoptical view of the effects of urinæmia, to recapitulate and condense the phenomena observed under the established morbid conditions of the affection.

1. *External Appearance of the Patient. (Early Stage.)*—Aspect, that of general feebleness; and, if the depraved state of the blood follow scarlatina or Bright's disease, a more marked pallor, than when other causes are operative in retaining the urea in the blood—together with a puffiness about the cheeks and eyelids. Generally, sallowness and anæmic hue, but sometimes blueness and congested appearance of the skin. More or less œdema of the extremities. Listless, confused, semi-idiotic manner.

2

(*Second Stage*).—Appearance that of a person apoplectically somnolent; degree, partial or complete; *modification*, by clonic contractions of the muscles.

(*Third Stage*).—Appearance that of one suffering from epileptic convulsions.

Either of the last two stages may be present singly, the other not occurring; or they may be combined and alternate.

2. *Disturbed Sensorial Manifestations. Early Stage.* ("Premonitory" of certain writers.)—Impaired vision; transient, partial and incomplete amaurosis (Reynolds, *op. cit.*); muscæ volitantes; tinnitus aurium; temporary deafness.

(*Later Stages*).—Deficient, and sometimes entire loss of sensibility; complete amaurosis; permanent deafness—the latter less common. Sensation is seriously impaired, but not very frequently wholly lost. Distinct cognizance of impressions not taken; but usual appreciation of injury to the corporeal surface, felt. (Reynolds, *et al.*)

3. *Motorial Manifestations.* (*Early Stage.*)—More or less severe *clonic* contractions of the muscles; heavy and unwilling motions; slight stertor, "even when the patient is awake." (Reynolds.)

(*Later Stages*).—Voluntary movements mainly absent; sometimes to be provoked by excitation; continuance of clonic spasms; epileptic convulsions, more or less strongly marked. Dr. Reynolds (*op. cit.*) remarks that the rigidity of the muscles observed during this period varies greatly, being sometimes excessive, "and much increased by movement of the limb."

4. *Mental Condition.* (*Early Stage*).—Listlessness; fretfulness; uneasiness; confusion of ideas, impairment of memory, or its entire loss; partial or complete, but light, delirium—noticed often during sleep, or "when falling asleep." (Reynolds, Braun, *et al.*)

(*Later Stages*).—The profound insensibility of true coma, but at first capable of dispersion—the patient can, by persistent efforts, be aroused; soon, merging of this state into that of complete and irrecoverable carus. Frerichs notices the fact that the usually mild delirium which may, but does not uniformly, attend this state, is characterized by reiteration of the same word for a long time. A maniacal state may follow the coma, when that disappears.[1]

The species of coma first referred to—whilst the patient can yet be aroused—very much resembles that arising from opium, or other narcotic poisons, acting with full force. Dr. Reynolds, referring to this fact (*op. cit.*, p. 109), says he has noticed this sort of coma in the great majority of urinæmic cases he has observed. He writes:—

---

[1] *See foot-note* 1, *page* 20.—"Some cases of puerperal mania, accompanied by albuminuria, and where no eclamptic attacks had occurred, are alluded to by Dr. Simpson."—*Duncan* (*note to Braun*, p. 136).

" The urinous element (whatever it may be) in the blood acts probably in a somewhat similar manner [*i. e.*, to that in which narcotics act]. There is not, however, in all cases of urinæmia, the notably contracted pupil that is observed in poisoning by opium. It is interesting to observe that the sensori-motor system appears to resemble, in its pathologic conditions, the spinal (or reflective) centre, rather than the cerebral (or intellectual). It is in a state of exalted rather than depressed activity, although both sensation and motion are severed from their purely cerebral relations (*i. e.*, from forming parts of perceptible and effective volition). There are several poisons which appear to act in a directly opposite manner upon cerebrum and spine (inducing at the same time coma and convulsions), but whether they contain different elements, whose action is thus separated, as Dr. Walshe once suggested, in a clinical lecture, the poison of urinæmia might be, I leave for future researches to decide."

5. *Special Functional and Organic Manifestations.*—In addition to the external appearances of the patient, as exhibiting deranged function of the skin, and perverted nutrition, the stomach and bowels may become excessively irritable. The vomited and other excreted matters, we are told by several observers, exhale ammonia when tested by hydrochloric acid; and the air expired from the lungs sometimes reacts similarly under the same agency (Frerichs, Johnson, Litzmann, Braun, *et al.*). The *pulse*, in the comatose state somewhat slow, rises, and is, at the same time, weak and irritable, in the convulsive periods.[1]

6. *State of the Urine.*—Confirmatory of the existence of obstruction to elimination and excretion. Depuratory processes at fault. The secretion is generally *acid* in reaction to tests, and *albuminous*—although cases of urinæmia occur in which albuminuria is not an element—casts of the *tubuli uriniferi*, and also blood-corpuscles and mucus-corpuscles are discovered by the microscope; and the urea is notably diminished in the specimens of urine passed. (Frerichs, Thudichum, Rees, Braun, Reynolds.)

A febrile condition, very similar to that of genuine typhus, is observed; and especially in connection with diminished excretion of the urine, or with its entire suppression. This is denominated by Frerichs, *febris urinosa;* the French writers designate it by the same term—" fièvre urineuse." There is delirium, excessive prostration, and a urinous odour pervading the excretions; and death is then imminent. Death may, in certain cases where the blood has been exceedingly impoverished and contaminated, follow epileptiform convulsions which are due simply to the deteriorated and devitalized blood. These convulsions should be distinguished from those arising from other causes. Sometimes, even in such cases, rupture of cerebral vessels may cause apoplectic coma, by effusion of blood.[2]

[1] It may be very much accelerated, and sometimes remains frequent throughout the affection.

[2] Drs. Watson, Todd, and George Johnson, have called attention pointedly to the fact of epileptiform convulsions springing from the circulation of impoverished blood in the cerebral vessels.

Should life be prolonged, and in cases where recovery is possible, and occurs, there may remain permanent injury to the general health; or special organs may be particularly acted upon. Hemiplegia, hemeralopia, amaurosis, and impaired mental vigour may be mentioned. Œdema of the lungs and serous effusion into the cavities are noticed—the cerebral ventricles, even, being invaded by an urea-bearing serum. Life is too frequently destroyed, however, to enable observers to enumerate many cases and facts bearing upon this portion of our subject.[1]

II. POST-MORTEM APPEARANCES.—It is universally conceded that very few structural changes, of consequence, and often *none whatever*, are found on necroscopic examination—after death from the mere action of uræmic poisoning—in the cerebro-spinal system. The apoplectic effusions and lacerations of the cerebral substance are only indirectly, if at all, referable to uræmia. The condition of this system is that which chiefly concerns us at this time; for we are seeking strictly for the effects, both vital and *post-mortem*, legitimately due to the toxic agent derived from the presence of urea in the blood. A full description, therefore, of what is found after death in the *kidneys*, does not seem to us pertinent to the question; for the usual and well-known renal lesions of Bright's disease are not the product, but simply the frequent cause, of the uræmic condition. We shall, therefore, endeavour to particularize the appearances presented after death by those organs which during life most strongly manifest the effects of the toxic agent, and with which it is most intimately brought into contact and relation—mentioning more succinctly such other concomitant lesions as have been observed. And in noticing the necroscopic phenomena in the cerebro-spinal system, we shall speak of such cases as have afforded the gravest indications of disturbance of the nervous centres during life, and especially decided coma and convulsions.

Brain.—Anæmic appearances, and a somewhat infiltrated condition, are noted; the consistence being most frequently diminished.[2] This state

---

[1] The *mania* which sometimes follows the comatose state in uræmic eclampsia, is generally well recovered from; but should not be confounded with that symptomatic of puerperal pyæmia.—Braun; who refers to Helm, Litzmann, "and others."

[2] Romberg (*Nervous Diseases of Man*, London, 1858, Syd. Soc. Ed.) when mentioning the results of cadaveric inspections in cases of *eclampsia parturientium*—which affection, it should be borne in mind, is now referred by such high authority to uræmic intoxication—says—rather adversely to what we find recorded by others—that in the cranial cavity we generally find considerable congestion, increased density of the cerebral tissue, plastic and sanguineous extravasations between the membranes, and in the ventricles, in the latter chiefly when apoplectic symptoms, a profound sopor, stertorous breathing, &c., have been associated with the convulsive affection." (*Loc. cit.*) The points in which this account chiefly differs from that given by Dr. Braun, are the frequency of congestion, and the increase

occurs irrespective of any abstraction of blood during life. The membranes of the brain are not commonly congested or "hyperæmic." Dr. Braun says, also, that inter-meningeal apoplexy is even more rarely observed than hyperæmia. He adds, that Helm and Kiwisch[1] very justly consider inter-meningeal apoplexy "as a secondary phenomenon produced by impeded circulation of the blood;" while it is looked upon by Litzmann as "a result of the uræmia."

Spinal Cord.—Examinations of the cord have been but infrequently made. (Romberg knew of *none* at the time of writing his *Treatise on Nervous Diseases*, the first edition of which was published in 1840 and the second in 1851.)

Braun states that Bluff, at one examination of the spinal cavity, "found much serum in it." Dr. Duncan reminds us, in this connection, that—as Dr. Christison first showed—the serum discovered in different regions of the body often has urea in it.

Dr. Todd, speaking of the condition of the cerebrum and spinal cord, after death following renal disease accompanied by coma and convulsions, says :—

"'The organic disturbance of the brain which accompanies and causes the comatose tendency, is, as I have already remarked, much less than the pulmonary affection. There we find nothing which the most zealous morbid anatomist could call inflammation; and, except the patient may have died in convulsions, we do not even find congestion—that most fertile of causes with a school of pathologists which is, I hope, fast disappearing. Indeed the brain is generally anæmic," &c. * * * (*Op. cit.*)

Dr. Simpson has reported (*Obstetric Works*, vol. i. p. 732) some cases of "puerperal convulsions connected with inflammation of the kidney," wherein effusion of blood and serum into the ventricles was discovered,

of the density of the cerebral substance, which the latter observer distinctly denies, as indeed do others. Romberg, whose whole description of the epileptiform convulsions of the parturient female is admirable, clear, and vivid, refers on the above points to Hauck (*Einiges aus dem Gebiete der praktischen Geburtshülfe;* in Casper's *Wochenschrift der gesammten Heilkunde,* 1833, vol. i. p. 133 ; and Velpeau, *Die Convulsionen der Schwangerschaft während und nach der Entbindung. Uebersetzt von Bluff:* Köln, 1835, p. 86). Romberg, as well as other authors, refers to the paucity of necroscopic facts connected with eclampsia puerperalis; and especially with regard to the spinal cord.* This is, even at the present day, true; but the latest observations, coming as they do from reliable sources, must be allowed the greatest weight. With regard to the appearances in the *brain,* then, a certain difference of statement exists between Romberg and other authors we have consulted. As to the *spinal cavity* and *the cord,* there are no new facts within our cognizance.

[1] Helm, Th., Med. Jahrbücher. Wein, 1839, bd. xx. s. 202 ; Kiwisch, Beiträge z. Geburtsk. Würzburg, 1846.—Braun, *op. cit.,* chap. VI.

* Never had been examined when Romberg wrote; but has been since.

with destruction of the right corpus striatum and outer portion of the optic thalamus, in one patient; together with encysted and degenerated kidney (morbus Brightii), and purulent-like matter, with adherent lymph, or false membrane, in another. In a third case, purulent-looking matter could be pressed out from the renal papillæ; no effused fibrin or coagulable lymph was discovered. The microscope did not decide the "whitish turbid fluid" to be pus—only epithelial cells, in large quantity, were found. The last patient had repeated convulsions, and died comatose, but no mention is made of any cerebro-spinal lesions being observed. The effusion of blood, and the laceration of cerebral substance in the first case related by Dr. Simpson, were truly apoplectic, and not referable to the intrinsic action of urea contained in the blood; although doubtless the amount present therein must have been considerable, since Bright's disease existed.

Dr. Watson (*op. supra cit.*, vol. i. p. 493), speaking of the appearances observed after death from apoplectic coma, after having referred to such a result from the action of retained urea, uses the following language :—

"On examining the brain, we may find a large quantity of extravasated *blood* spread over its surface, or lying within its broken substance; or a considerable effusion of *serous fluid* collected within its ventricles; or we may detect *no* deviation whatever from the healthy structure and natural appearance of the organ. The congestive pressure (if, indeed, it existed) has left no prints of its action."

The following observations, from the *Manual of Pathological Anatomy*, by Drs. Jones and Sieveking, are exceedingly apposite in this connection :—

"The researches of Bright, Frerichs, and others have demonstrated the close relation of the state of the blood to cerebral disease; and science has shown, what, previously, was purely hypothetical, that the most fatal conditions may be thus induced without any palpable changes being wrought in the cerebral tissues. It does not, however, follow, that because we see no changes, none have taken place. The poison that we know to be in the blood may elude our chemical tests, and yet cause death. Then, seeing how limited our knowledge of the nervous system is, it is not to be wondered at that, although the manifestation of altered function is so great as to force the belief in its altered constitution, it is not in our power to prove the latter to the perception; but, as Dr. Watson remarks, 'whatever may be the nature of the unknown, and, perhaps, fugitive physical conditions of the nervous centres, thus capable of disturbing, or abolishing their functions, it is useful to keep in our minds a distinct and clear conception of the fact, that there must be some such physical conditions.'"

Mere uræmia, therefore—presuming the occurrence of convulsions and coma—it would appear, leaves the brain anæmic in appearance, and possibly somewhat softened (more dense, according to Romberg; refer to page 20); the more decided and destructive appearances are owing to rupture of vessels, and consequent extravasation, with its consequences, and to dilatation of the ventricles with serum.

Lungs.—These organs are constantly found in an œdematous condition, and sometimes emphysematous. Dr. Braun, recording the fact that emphysema was long since observed by Böer, says that it is now considered "as always the secondary result of the fits"—*i. e.*, uræmic convulsions. *Op. cit.*, p. 62.

Heart.—This organ is reported to be usually "empty and flaccid." (*Braun.*) We may thence infer feebleness of circulation, and impairment of its own tonicity, and of that of the bloodvessels, by reason of the impoverished state of the blood, their natural stimulant.

Kidneys.—Generally, and according to some authors, always, the kidneys exhibit more or less extensive and advanced signs of Bright's disease. In cases where the retention of urea in the blood has been caused by some other agency—such as obstruction, etc., there would naturally be traces of congestion, and perhaps of inflammation, although not uniformly; a nearly natural state might well enough exist.

As we previously intimated, it does not seem necessary to give, in this place, an elaborate account of the changes wrought by granular and fatty degeneration of the kidney, in connection with uræmia; and for reasons already stated. Moreover, these general appearances are well known, and abundantly set forth in many treatises. Those who would see, however, an admirable and somewhat condensed account of the changes of this nature effected in the renal tissue in urinæmic cases connected with pregnancy (*eclampsia puerperalis, seu gravidarum*), should consult the chapter of Dr. Braun's work, to which we have so frequently referred. A few extracts only will be made by us, and those chiefly to call attention to certain prominent points in the renal necroscopic phenomena.

Dr. Braun bases his descriptions on the three forms of Bright's disease proposed by Frerichs.

In the first stage, that of hyperæmia and commencing exudation, the surface of the kidney is smooth, the capsule is easily removed, the plexus of veins on the surface of the kidneys is dilated, and full of dark blood." (*Loc. cit.*, p. 62.)

"The pyramidal masses [renal papillæ] are likewise hyperæmic, and their injection is striped. The mucous membrane of the pelves and infundibula is swollen, and covered with vascular arborescence; and they contain a bloody fluid. Apart from hyperæmia, the finer structures of the kidneys do not appear to be essentially injured. Hemorrhagic effusions are very frequently observed, which sometimes take their rise from the glomeruli; sometimes from the vascular plexus of the tubuli uriniferi, sometimes from the veins on the surface of the cortical substance." (p. 63.) In the first stage, the epithelial lining of the uriniferous tubes is stated not to be essentially altered; the tubes themselves are often filled with exuded blood— "fibrin-cylinders."

In the second, or exudative stage, fatty degeneration commences and progresses, the kidneys becoming large, and heavier than they are normally. The capsule of the kidney can be easily separated. The pyramidal masses are dark red. The infundibula have a dirty-red mucous surface. The glomeruli (vascular knots, Malpighian corpuscles), which may be drawn out with a curved pin, are covered with a fine granular matter, and partly with solitary or grouped fatty corpuscles, which, by the addition of acetic acid, become transparent.[1] Between the glomerulus and the capsule lies a thick stratum of firm exudation, of granular structure, and having fat droplets, and sometimes crystals of cholesterine.

The contents of the epithelial cells of the tubuli uriniferi next become degenerated—they are "decomposed in aggregations of granules."

Fig. 1.

Malpighian Corpuscle, from the kidney of a patient who died of *Eclampsia Parturientium*.— (From Wedl's Pathological Histology, p. 260.) "The surface is covered, partly with a fine granular substance, partly with solitary and aggregated fat-globules, which were not further changed by acetic acid or carbonate of soda. * *"—(*Loc. cit.*)

In the third stage, that of retrogradation, atrophy of the kidney is progressively induced. The pyramids of Malpighi and Ferrein are observed to be less atrophied than the cortical layer. At their bases, granulations occur between the straight tubuli, and press the latter apart.

The width of the renal pelves is increased, and their lining mucous membrane swollen and covered with a network of "varicose vessels, of an uniform grayish-blue colour."

"In those who die of uræmic eclampsia during pregnancy, atrophy of the kidneys is less frequently observed than the first two stages of Bright's disease." (*Op. cit.*, p. 69.)

III. URÆMIA IN CONNECTION WITH PREGNANCY AND PARTURITION.—If we receive the opinions of certain modern pathologists as true, the retention of urea in the blood of the pregnant female is productive, both to herself

---

[1] Wedl mentions that acetic acid rendered the fat-globules more distinct.

and to the fœtus, of the most disastrous results. The views of Dr. Braun, already referred to, have been adopted by a large number of able and practical men in our profession; and, since the publication of his volume—and mainly, in Great Britain and this country, through Dr. Duncan's translation of the chapter on uræmic eclampsia—have elicited great attention, and will doubtless prove an incitement to yet more extended and close observation.

Whilst many have joined Dr. Braun in this new theory, there were at the time his work appeared, and probably still are, several distinguished names arrayed against it. These are all mentioned by the author himself, who also states the objections made by them, and gives what he considers the refutations thereof. Dr. Braun's belief, as declared in his treatise, is simply this: *that the convulsions caused by uræmic intoxication in acute Bright's disease, and puerperal eclampsia, are identical.* This proposition, as he tells us, has been energetically defended by Frerichs, Litzmann, Wieger, Oppolzer, himself, and many others; but it has been assailed by Marchal, Siebert, Depaul, Legroux, L'Huillier, Stoltz, Seyfert, Levy, in very valuable articles, and also by Scanzoni. With much anxiety these observers have tried to prove that the Brightian degenerations of the kidneys, which, it cannot be denied, are found in the bodies of those who have died of eclampsia, are consequences merely of the convulsions—only accidental, secondary phenomena of the hyperæmia caused by the eclampsia, and of hydræmia (*plethora serosa*).

Scanzoni, whose arguments against Dr. Braun's views are summed up and given by the latter author, refers the true *eclampsia parturientium* to an "irritability of the motor system of nerves which has been induced by pregnancy, and increased by the act of delivery."

The arguments *pro* and *con* have been actively carried on; and Scanzoni's conclusions were replied to by Wieger in June, 1854, and by Litzmann in 1855. In addition, the industry and zeal of Dr. Braun have enabled him to collect a truly imposing array of facts from *post-mortem* observations made by himself and other reliable practitioners, and in the majority of which abundant evidence of Brightian renal lesions existed. In some of the cases, where hyperæmia of the kidneys was found, microscopical examination was prevented by "accidental obstacles;" but the author does not consider this "any proof of the absence of Bright's renal exudation." Dr. Braun, while setting forth these investigations, says that Wedl[1] explains the non-discovery of fatty degeneration in the kidneys in several instances of death from eclampsia, "by the fact that in many cases a dissolution of the Malpighian bodies is effected by the fluid exudation, and hence in every diffuse inflammation of the kidneys an evident fat-metamorphosis of the contents of the Malpighian capsules does not ensue."[2]

---

[1] Grundzüge der pathol. Histologie, Wien, 1854, S. 306.
[2] Loc. cit., pp. 74, 75.

Hasse, according to Dr. Braun, has never seen eclampsia puerperalis without Bright's disease. The necroscopic and microscopic observations of Wedl, Gustav Braun, Lumpe, Hecker, Devilliers, Regnauld, Simpson, Blot, Cahen, Wieger, Litzmann, Credè, Sabatier, and Hohl have contributed to establish the theory propounded in Dr. Braun's volume.

In reference to the *etiology* of "eclampsia parturientium" ("uræmic eclampsia" of Braun), Romberg[1] only hints at the agency which Braun has distinctly announced as, in his opinion, the chief efficient cause. Thus, while the latter ably demonstrates his views as to the retention of urea, and the accompanying albuminuria, the former refers to "retention of urine owing to pressure of the gravid uterus upon the bladder;" and afterwards adds: "Future investigations must determine whether albuminuria, which often supervenes during pregnancy, may, when fully developed during the last months, possess any etiological influence." The "investigations" of Braun and others certainly seem, if not already to have determined this point, likely to lead shortly to a satisfactory settlement of the whole question.

Supposing, then, in conformity with the new doctrines thus announced, that the convulsive attacks and intervening coma observed in puerperal patients are owing to toxæmia by retained urea, let us examine the collateral results of such a condition—first, in regard to the mother; and next, as respects the fœtus. Having already detailed the general effects produced upon the system by uræmic intoxication, we may properly direct our notice, at present, to the influence exerted upon the *puerperal state*, and upon the *life* of the mother and that of the fœtus.

Referring to our previous enumeration of the results, both vital and *post-mortem*, which are observed after genuine uræmic convulsions, we would only add thereto that the *uterus and its appendages* are generally found healthy after eclampsia puerperalis, or, at all events, do not necessarily deviate from their usual condition after labour has terminated; unless, of course, there has been some pre-existent or concomitant disease in, or alteration of, the organs. The infrequency of metritis and peritonitis is mentioned by some observers;[2] but Churchill[3] speaks of the great tendency to abdominal inflammation after the labour is over, and quotes Denman, who first mentioned this, and Collins and others, as confirming it. Braun also refers to the danger of puerperal affections coming on after eclampsia, especially if "an epidemic of zymotic diseases prevails."

The *spleen* is said to exhibit "the large dimensions it possesses in pregnancy and childbed."[4]

[1] A Manual of the Nervous Diseases of Man ; Sydenham Society's edition, London, 1853, vol. ii. p. 189.
[2] Romberg, Velpeau, *et al.*   [3] Theory and Practice of Midwifery.
[4] Braun, op. cit., p. 62.

We now proceed to the consideration of the special influences exercised by uræmic intoxication during the puerperal state.

1. *Influence of Uræmic Eclampsia on the Duration of Pregnancy.*— Uræmic eclampsia is generally sudden in its accession, prompt in its results; and the testimony of those who have had the most experience in regard to it, is that it either causes death rapidly, or else that it is completely and with considerable celerity recovered from. Long consequent illness and sequelæ, *from the eclamptic state merely*, are not common. The true uræmic eclampsia occurs in the majority of cases in the latter part of the period of pregnancy. It may, also, only appear at the time of the labour itself; after the child is born, and before the after-birth has been delivered; or, finally, during child-bed. When supervening within the latter half of utero-gestation, premature labour is the result—as a rule.[1]

The latter two months of pregnancy are stated to be peculiarly obnoxious to the occurrence of convulsions. (Churchill, Romberg.) The *cause* of premature labour is either excitation of uterine contractions by the power of the abnormal action going on in the nervous centres,[2] or it may be, partially at least, ascribed to the presence of a dead fœtus, whose destruction is referable to the eclamptic condition. Ramsbotham, we observe, does not think the latter occurrence a cause. Churchill queries whether it may not be such; and the supposition is at least plausible. Dr. Copland (*Dict. of Med.*) speaks of cases where "the child has been unexpectedly born during the violence of the convulsions, as if expelled by them with unwonted celerity." Again, he states that the worst forms of the attack are often attended by a firm, spasmodic constriction of the cervix uteri, preventing the expulsion of the fœtus.

2. *Influence of Uræmic Eclampsia upon the Life of the Mother.*— Although puerperal convulsions—we here use the term, let it be remembered, as synonymous with uræmic eclampsia—are comparatively a rare affection, yet they make up for the element of infrequency of occurrence, by violence of manifestation and an alarming ratio of fatality. As the liability to an attack is greater towards the last of the period of utero-gestation, so, generally speaking, is the danger to the life, both of mother and child, at that time. Not only is this true if we merely refer the eclamptic condition to the increased sources of irritability to which the nervous system has

[1] It is well known that Bright's disease, without any other influence, will cause premature labour; Braun says this is true in about 25 per cent. of cases. Add *convulsions*, and the danger is manifestly increased.

[2] Blundell (Principles and Practice of Midwifery, Am. edit., p. 418), referring to this point, says: "Sooner or later, * * if the fit continue, parturition commences of itself, without the interference of the accoucheur; and * * a sudden emersion of the fœtus occurs." He also refers to the occurrence of delivery during convulsions, unknown to the attendants.

become liable, by reason of the pregnant state and its advanced stage; but also it is easy to see that a poisoned blood is all the more likely to act with increased morbid force—especially upon fœtal life. Again, the danger from convulsions diminishes according as they approximate to the term of delivery; and we are told, by a most competent authority, that the fits diminish in force in 31 *per cent.* of the cases, cease entirely in 37 *per cent.*, and only continue of the same intensity in 32 *per cent.*, *after the uterus is evacuated.* (Braun.)

Dr. Braun refers to 15 deaths out of 45 cases, of which he has published accounts—being exactly *one-third.* He speaks of nine cases occurring in his practice within the last three years, all of which terminated in recovery. Thirty *per cent.* of the cases, it is estimated, prove fatal. Romberg gives a higher ratio of mortality. His statement is, that above one-half of the women attacked, died within twelve, twenty-four, or thirty-six hours. Churchill finds from his statistical investigations, that the proportion of fatal cases is above one-fourth. He intimates that there has been a tendency, of late, to diminution in the mortality-rate, which, at one time, he asserts, was very much larger. It has not been thought that much, if any difference as to fatality can be ascribed to early or late supervention of the affection; at all events, lateness of attack has not been allowed more weight, in this respect, by observers, when it is a first case. Some patients have several attacks in successive pregnancies, and finally die in one. A reiteration, therefore, of the accident, must be deemed of unfavourable augury. The concomitance of coma, with apoplectic stertor, and the approximation of the fits, so that they become, as is sometimes the case, almost continuous—and especially when these conditions obtain in plethoric and strongly constituted patients—are, almost without exception, fatal elements.

Recovery from the mere eclampsia may take place; and there may be some extensive dropsical effusion, some injury to the brain or spinal cord, or a permanent and increasing œdema of the lungs, disease of the heart, etc., which will compromise life at a later period; but, as a rule, if the patient escape the eclampsia, and its *immediate* results, recovery is usually not tardy, and, moreover, is complete. The occurrence of rupture of the uterus, in itself sufficiently grave at any time, we need scarcely say, very greatly aggravates the unfavourable prognosis attaching to eclampsia. Dr. Copland remarks (*Dict. of Practical Medicine*), that puerperal convulsions "should never be considered devoid of danger, more especially when they occur after delivery, or in consequence of great exhaustion of vital power, or of uterine hemorrhage. When they are slight, are unattended by stertorous breathing, or by paralytic or apoplectic symptoms, and when parturition is so far advanced as to readily admit of its completion by art, less danger may be feared."

The balance of prognostic opinion, it will be seen, is against recovery ; and Blundell speaks of *post-partum* convulsions as being the most dangerous.

3. *Influence of Eclampsia on the Life of the Fœtus.*—We have already intimated the danger arising from eclampsia parturientium to the life of the fœtus. Relatively, it is even greater than that threatening the mother. In referring to this point, those who have had the largest experience, use such expressions as the following : "In almost all cases the child is still-born, often putrid." (Churchill, *op. cit.*) "The life of the fœtus is endangered so long as it is nourished by the uræmic blood of the mother." (Braun, *op. cit.*) "The infant almost invariably dies when the disease occurs during the last months of pregnancy; it may be saved when the eclampsia supervenes during parturition." (Romberg, *op. cit.*, p. 189.) "The infant is generally, though by no means universally, born dead, when the woman has been the subject of convulsive seizures, especially if the attack has occurred early in the labour, and continued for any length of time." (F. H. Ramsbotham, *Principles of Obstetric Medicine and Surgery.*) The latter author pointedly and happily refers to *toxæmic action* on the blood of the fœtus as the most likely cause of its death *in utero ;* and mentions a case from Spence, where the child being removed alive by Cæsarean section from a mother just dead from convulsions, died itself, in convulsive paroxysms, in less than an hour. The latter statement leaves us to presume a poisoned condition of the blood. M. Ménard states, that, in the majority of cases of death by convulsions, previous to delivery, the child has been found dead ; the contraction of the features and extremities showing that it had participated in the affection of the mother. Dr. Copland, who notes this remark (*Dict. of Med.*), says that it "wants confirmation."

We thus see that the likelihood of the life of the fœtus being maintained after eclampsia has been declared in the mother, is extremely small, as might, indeed, be expected ; and if the child be born alive, there is great probability that it may either not survive long, or that it will be more or less morbidly affected by the accidents occurring during its intra-uterine existence.

Hereditary transmission of eclampsia, uræmia, or Bright's disease of the kidney, to a suckling, says Dr. Braun, "has not yet been demonstrated, and only Simpson has found albuminuria in a suckling born of an eclamptic mother." Dr. Duncan, in a foot-note to the paragraph from part of which the last remark is taken, says that "if the uræmia persists in a nursing woman, urea may be present in the milk, as has been shown by several observers, and may disturb the health of the suckling."

URIC ACID.—(Symbol: $\overline{U}$.   Formula: $C_5HN_2O_3+HO$.)[1]

Synonymes.—Lithic acid; Urylic acid.—PROUT, BIRD, *et al.*

This substance is the next constituent of the urine, in the order of enumeration we follow; and the effects of its retention in the blood will therefore now engage our attention.

It is recognized as a constant ingredient of healthy urine, and holds a very intimate relation to urea. (Thudichum, *op. cit.*) "It forms less than $\frac{1}{2000}$ part of the urine in man; but its proportion varies much in different animals." (Dr. George Johnson, *op. cit.*, p. 50.) In the blood, it is always found in union with an alkaline base; and it is not soluble in that liquid. It appears in the blood in the form of urate of soda, or of urate of ammonia. (Liebig, Simon, Thudichum.) Dr. Thudichum states that the urates are always acid salts—*i. e.*, that there is excess of uric acid—and he has advanced some ingenious reasons for denying the accuracy of Dr. Golding Bird's theory and explanation of the secretion of uric acid in the form of urates. With this chemical discussion we have no concern; the indication of the pathological states induced by retention of the acid in the blood, in the form above specified, being now our object. Before approaching the subject in detail, however, it does not seem particularly out of place to allude to a practical remark by Dr. Thudichum (*op. cit.*, p. 100) upon the deposition of the urates from the urine. Researches and observations in reference to this point, cannot fail to be of importance in the treatment of such cases. Dr. Thudichum says :—

"If the presence of a deposit of urates be taken as an indication of the saturation of urine by these salts, and if the latter be assumed ordinarily to be of the usual amount, deposits of that kind become more valuable as signs of a diminished secretion of water by the kidneys than of any other symptom. As the appearance of a deposit of urates is always accompanied by morbid sensations and objective symptoms—in the healthy by thirst at least, if by nothing more— the conclusion is simple enough. *The individual whose urine has deposited the urates does not drink water enough, and must drink more, and must drink so much that the urine, at the ordinary temperature of the air, shall remain clear.* Of course, in some cases this will be neither possible or advisable; but in most cases of acute and febrile disease it should be a plan of treatment. I have certainly seen it attended by beneficial results in many cases; I have also observed the contrary—want of water in the system—to be a source of disease."

[1] Thudichum.

Chemical Composition of Uric Acid.

THUDICHUM.		G. BIRD.
Carbon	35.714	$C_{10}, N_4, H_4, O_6, C_2, H_4, N_2, O_2 + 2C_4, NO_2 = 168$
Hydrogen	1.191	
Nitrogen	33.333	
Oxygen	19.048	
Water	10.714	
	100.000	

GOUT.—Whenever, from failure to eliminate the uric acid from the blood, that substance accumulates therein, the abnormal effects of its presence do not long delay their manifestation.[1] It is well known that the ingestion of large quantities of highly-azotized food, and a rich diet generally, together with the free use of malt liquors and of acid wines—Madeira amongst others—is productive of an abnormal increase of uric acid in the system; and, consequently, luxurious livers have long constituted the majority of sufferers from gout—a malady which abundantly declares its *fons et origo*, by the tendency to abundant deposition of the *urate of soda* in different parts of the body; its seat of election being the smaller joints—as those of the toes and fingers. In the latter, the deposit is often very plentiful. We have seen not long since—and the occurrence may not be very uncommon—a patient who could write his name with his "chalk-stone" knuckles; a woful example, truly, of diverted and arrested excretion![2]

As has been previously mentioned, Dr. Garrod, of London, first called attention, not only to the presence of uric acid in healthy human blood, but also pointed out the fact of its abnormal amount in connection with gout. He did not, however, then wish to be distinctly understood as declaring the acid the sole *materies morbi* in that disease, as may be seen by his remarks in a "Postscript" to his highly interesting paper, communicated to the Medico-Chirurgical Society, upon the subject. This important contribution to scientific medicine was read before the Society,[3] February 8, 1848, and the postscript just alluded to bears date July 26, 1848. (Vide *Medico-Chirurgical Transactions*, vol. xxxi.)

Daily observation tends to show that the relation of cause and effect may more and more safely be predicated in regard to the presence of an excess

---

[1] Bernard recognizes the accumulation of uric acid in the blood, either by arrest of the renal function, causing its elimination to cease, or by an exaggerated production of it, as in gout. (*Liquides de l'Organisme.*)

[2] An instance of this is related by Dr. Watson (*op. cit.*), and Dr. Todd has alluded to the condition. After all, cases of this extreme nature may not be so common as we have intimated.

Dr. Garrod (*On the Treatment of Gout and Rheumatic Gout*) remarks upon this point: "Comparatively few gouty patients become the subjects of visible chalk-stones, at least to any extent, or such as to induce deformity; at the same time, I am convinced that their occurrence in a slight degree is by no means so rare as has been generally assumed. Sir C. Scudamore stated, that in 500 cases of gout he only found them 45 times, or in less than 10 per cent. From my own experience, I consider these numbers far below the real proportion, being confident that their existence is frequently overlooked, as they are sometimes deposited in parts of the body scarcely to be expected." Dr. Garrod thinks that gouty concretions are more frequent upon the cartilages of the ear than anywhere else; contrary to what has usually been recorded. He refers to the *Medico-Chirurgical Transactions*, vol. xxxvii., 1845, pp. 74, 75.

[3] By Dr. C. J. B. Williams, upon some of whose patients Dr. Garrod's experiments were made.

of uric acid in the blood and the phenomena of gout. Writers upon the subject, both near the time of Dr. Garrod's first researches, and later, have varied somewhat as to the completeness with which they have expressed themselves in respect to establishing uric acid as the active agency in gout. The majority of testimony seems to be affirmative. Dr. Watson (*Principles and Practice of Medicine*, 1848) seems to have then regarded the morbid agent as recognized. We find, indeed, in the edition of Dr. Carpenter's *Physiology* published in 1846, very positive language as to the conspicuousness of uric acid in gouty affections; he says : "When it [uric acid] is imperfectly eliminated, we are assured of its accumulation in the circulating fluid, by its deposition, in combination with soda, in the neighbourhood of the joints—forming gouty concretions, or chalk-stones." He thus appeared to recognize the cause of the diseased condition as lithic acid. There are those, however, who, even at the present day, speak with less distinctness as to an excess of uric acid being the sole and sufficient *materies morbi*. Thus, Dr. Barclay (*A Manual of Medical Diagnosis*, London, 1857) writes : "The researches of recent times have gradually led to the discovery of an important element in gout—the presence of an excess of uric acid in the blood. This knowledge holds out a prospect of our arriving ultimately at more accurate diagnosis; at present, it is only in the hands of a few that such a chemical test can be relied on." The opinion is a very guarded one—decidedly non-committal ; we think more influence than it implies may safely be allowed to the "element" in question.

It will, at least, not be disputed that gout is a blood-disease. Amongst many other observers who might be cited on this point, we select Dr. George Johnson, as furnishing comprehensive testimony. Referring to gout as a cause of renal disease, he says : "It would be useless to occupy the time of my readers by lengthened arguments to prove that gout is a blood-disease, since all the phenomena of the disease clearly indicate such an origin, and can be explained on no other supposition." (*Op. cit.*, p. 78.) He then alludes to the intimate connection between gout and the uric acid diathesis. Thus, then, when such a diathesis prevails, or when, by some obstructing agency, the elements of the urinary secretion are retained and accumulate in the blood, the gouty accidents, amongst others, prevail. If uric acid be prominent, the corresponding series of symptoms seems as sure to occur, as does that following the retention of urea when *that* substance is retained, in excess, in the circulation.[1] If the uric acid, therefore, is received as the true *materies morbi* in gout, we have, at once, the following easily-deduced sequence of morbid effects :—

First, deficient depuratory action ; next, accumulation of uric acid in

---

[1] The blood contains, as Dr. Garrod remarks, "mere traces" of uric acid *in health*. This fact, however, in no degree invalidates the agency of the acid as *a causative element in gout*.

the blood; *dyscrasia,* chemical and physiological; deposition of urate of soda in various parts; the objective phenomena of gout; the *sequelæ* of gout—amongst others, as Dr. Johnson points out, renal disease.   Under the light of modern pathology, we do not think the above any too weighty a burden to lay to the charge of retention and accumulation of uric acid in the blood.

There may remain a certain quantity of uric acid in the urine, at the same time that the analysis of the blood shows an abnormal amount therein. This would indicate a large supply from some source—either from waste of the tissues, or from the excessive ingestion of highly-azotized food, to which latter cause we some time since alluded.   Dr. Copland notices, in a comprehensive and satisfactory manner, the "Pathological Relations of Uric Acid and Urate of Ammonia" (*Dict. of Pract. Med.*); and M. Becquerel (*Séméiotique des Urines*) agrees with him in his views.   Copland, referring to this, considers the chemical explanations offered by Liebig, in connection with certain of these pathological points, "opposed to clinical observation."[1]

Dyspepsia, with mal-assimilation of food and consequent deficient nutrition, or arrested cutaneous function and habitual costiveness, no less than obstructed excretion of the urine, may engender an excess of uric acid, and the latter may be retained in the blood. (Copland, *loc. cit.;* Chambers, *Digestion and its Derangements; et al.*)   It is likewise true, conversely, that in depraved, deteriorated, and anæmiated states of the circulation, uric acid is diminished in the urine.   If this ill-adjusted balance implies the throwing of an unusual quantity of the substance into the blood, the latter circumstance may have no small amount of influence, if not in causing, at least in continuing, the disorder of the blood itself and of the constitution generally.   Gout occurring under such circumstances—as it is not very infrequently known to do—would be appropriately termed *atonic gout,* or what Dr. Todd (*Clinical Lectures*) terms "*asthenic gout,*" in contradistinction to his "sthenic gout," where the affection occurs in robust, well-constituted, and richly-nourished individuals.[2]

In reference to the dyspeptic condition of gouty patients, often so exceedingly troublesome, Dr. Chambers (*op. cit.,* American edition, 1856, pp. 294-95) refers rather scornfully to the influence of uric acid as a noxious element.   He is remarking upon the tendency of the food to become acid after ingestion, and to lie unchymified—not "passing onwards."   This

---

[1] Dr. Copland does not believe that the presence of urate of soda in the blood of gouty patients precludes "the elaboration of a portion of the uric acid and its compounds, or the modification and metamorphosis of one or more of their elements by the kidneys." (*Dict.*)

[2] These two forms of gout, we conclude, are those termed by Dr. Druitt "*high*" and "*low*" gout. (*The Surgeon's Vade Mecum,* 1859.)

3

state of things, so common in hereditarily gouty persons, Dr. Chambers is inclined to explain by considering its pathology " to be a slight flux of mucus, deficient gastric secretion, and yet a vigorous, sometimes even excessive appetite. Hence, they have not that check of failing desire for food which makes the meals of other invalids moderate, and eat more than their imperfect gastric juice can digest. This is a simpler, and, therefore, more probable explanation than the usual chemical talk about uric acid, &c., which might be substituted for it." It seems there is some fault in the working of the hidden *chemistry* of the body, however; and, although Dr. Chambers's explanation is doubtless very correct in reference to the influences and agencies of which he speaks, yet the overt action and manifestations of excess of uric acid, in the visible form of urate of soda concretions, sufficiently show the importance of the part it plays in the *tout ensemble* of gouty affections.

The original conclusions of Dr. Garrod, given in the admirable paper to which we have referred, were these :—

" 1st. The blood in gout contains *uric acid* in the form of urate of soda, which salt can be obtained from it in a crystalline state.

" 2d. The uric acid is diminished in the urine immediately before the gouty paroxysm.

" 3d. In patients subject to chronic gout with tophaceous deposits, the uric acid is always present in the blood and deficient in the urine, both absolutely and relatively to the other organic matters, and the chalk-like deposits appear to depend on an action in and around the joints, &c., vicarious of the ' uric-acid-excreting' function of the kidneys.

" 4th. The blood in gout sometimes yields a small portion of urea (no albumen being present in the urine)."

These conclusions were all duly sustained by analyses and experiments upon patients in University College Hospital; and the record of these demonstrations is at once satisfactory and highly interesting.

In respect to the supervention of gout in low and debilitated states of the system, to which allusion has previously been made, Dr. Garrod very clearly explains " two facts" then considered opposed to referring the pathology of gout to the humoral doctrines. We may say, *en passant,* that the humoral pathology seems the only reasonable mode of explaining the affection, and that, as may be distinctly perceived, it is coming more into favour of late, than it has been for a long time, in respect to many diseased conditions.

Dr. Garrod's remarks on the above point are :—

" Any undue *formation* of this compound (urate of soda) would favour the occurrence of the disease ; and hence the connection between gout and uric-acid gravel and calculi ; and hence the influence of high living, wine, porter, want of exercise, &c., in inducing it." Then, speaking of the "two facts"—viz., *hereditariness* and the supervention of gout in *low states* of the system—he says :

" We can understand that the peculiarity of the kidney, with reference to the excretion of uric acid, may be transmitted; and likewise that, when the function in question is permanently injured [viz., the 'uric-acid-excreting' function], it will not require an excessive formation of the acid to cause its accumulation in the blood." (*Loc. cit.*, pp. 93, 94.)

Dr. C. J. B. Williams (*Principles of Medicine*), referring to the fact that gout had been generally admitted, by inference, to depend on the existence of an excess of uric acid in the system, chronicles Dr. Garrod's experiments and analyses, which, as we have stated, were first made on patients under Dr. W.'s care, in hospital. He says :—

" Gout, and the commonest kind of urinary gravel are now generally considered to depend on the production in the system of an excess of lithic acid." (*Loc. cit.*) The case from experiments upon which Dr. Garrod was first enabled to draw his reliable conclusions, "was one of chronic gout; and further illustrated the pathology of the disease, by a total absence of lithic acid in the urine, until during the exhibition of colchicum, when its characteristic crystals appeared under the microscope."

Sufficient testimony, it would appear, has been adduced, to render the position tenable which ascribes the gouty paroxysm to an excess of lithic acid, circulating in the blood, and finally deposited in various parts of the body.[1] It does not seem to devolve upon us to describe the phenomena of a fit of the gout—expected as we are, merely to signalize the " effects" of retention of the various elements of the urinary secretion in the blood, we should strictly confine ourselves to such specification, and to the adduction of the best evidence afforded in its support.

In the first place, then, we may refer the phenomena of gout, more or less completely, to that disorder of the eliminating function of the kidneys, which permits the latter organs to refuse the excretion of uric acid. The latter substance is then necessarily thrown into the circulation, and its tendency—under the circumstances—is to accumulation in the blood. As it accumulates, it is converted, most commonly, into urate of soda, and the deposition of the latter substance upon and into various tissues of the body is a *quasi* vicarious discharge of the uric acid, not excreted by its legitimate channels, the kidneys. This condition is accompanied by the objective phenomena of gout, viz., pain, of an exquisitely acute and torturing character ; feverishness, dyspeptic symptoms, and general *malaise;*[2] at the close of the paroxysm, tense, shining, and often excessive swelling of the affected part; finally, entire remission of the symptoms, and better health than before the attack—owing, of course, to the elimination of the *mate-*

[1] See Appendix, Note B.

[2] " An impure state of the blood, arising principally from the presence of urate of soda, is the probable cause of the disturbances which not unfrequently precede the seizure, and of many of the anomalous symptoms to which gouty subjects are liable." (A. B. Garrod, *op. cit.*, p. 341.)

ries morbi. The *subsequent course* of things will depend very materially upon the habits of the patient, and upon the power he has of restraining his appetites; the fact of hereditary predisposition, or the contrary; and whether the management of the initial attack is judicious, or otherwise. By renewed attacks—chronic gout—the system of course becomes more shattered, less capable of resistance, and less amenable, too often, to remedies. Locally, a variable, often an excessive, amount of injury supervenes. It is at these periods that the lithate of soda—the morbific material—is deposited in various places—principally about the smaller joints. Nature, in her efforts to eliminate this material by other channels than those which are the legitimate ones, does the best she can, but often terribly cripples the subjects of this trying affection (see Fig. 2). Connected with this chronic form of the complaint, we are apt to notice the most troublesome combinations of dyspeptic ailments; and in these the condition of the patient becomes the most unpromising possible, both from the local and from the systemic difficulty. As Dr. Williams states, the uric acid, in such cases, "seems to be engendered in great abundance, and although thrown off in large quantities in the urine for an indefinite period, yet never leaves the body free. Such cases are commonly either hereditary, or those which have been rendered inveterate by intemperate habits, or neglect of proper treatment." (*Op. cit.*)

Fig. 2.

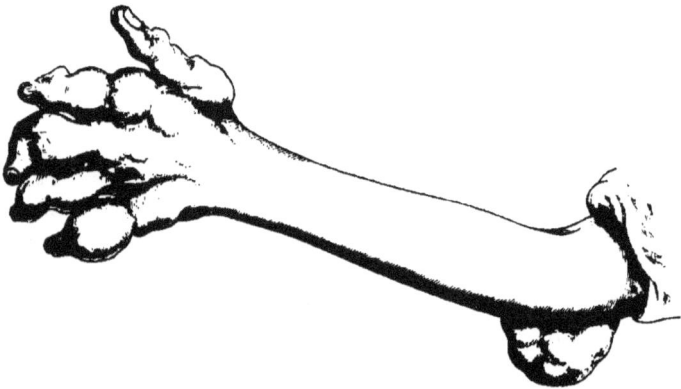

Elbow and Hand deformed by Gouty Enlargements.—From Dr. Garrod's work (Nature and Treatment of Gout.)

Sequelæ of Gout.—Considering gout as one of the affections ascribable to the retention of uric acid in the blood, let us inquire what are the principal subsequent results. These may, of course, be properly referred to the same cause, as "morbid effects."

We have seen that symptoms of general febrile disturbance, which are of course accompanied by nervous apprehension, fretfulness, heat of skin, more or less sleeplessness, and scanty urine—and which at last is loaded with the lithates, whenever the paroxysm of gout comes on, and at its decline, especially—with hypochondriacism, cramps, flatulence, diarrhœa, but more commonly costiveness (Watson), and more or less general and indescribable *malaise*, both precede and accompany gouty attacks. The dyspeptic symptoms may long remain; although, with care, they may be made sooner or later to disappear. .

One of the most serious sequelæ of gout may be considered that wherein the kidneys are affected. Perhaps we may best describe the state we wish to indicate, by terming it the result of a sort of *recoil* upon the kidneys, of the antecedent morbid action in the economy. This is of course entirely the opposite condition of that obtaining when the kidneys fail to eliminate the uric acid, *i. e.*, when their "uric-acid-excreting function" is suspended. In the latter condition of things, we have the blood-disease, the results of which we have sketched above; but when the uric acid begins again to be excreted by its natural passages, the kidneys are very likely to be more or less irritated in the process. There may then be violent pain (*nephralgia*), and nephritis or true inflammation of the substance of the organs may occur. Fully as unfortunate is that state when there is such an amount of uric acid as to be thrown down from the fluid part of the urine—the latter not being sufficient to hold it in solution, and thus carry it out of the body—when it is exceedingly apt to become concreted in various parts of the urinary passages—thus producing obstructions more or less seriously endangering health, and vitiating, in different degrees, the integrity and usefulness of the organs involved. Permanent irritation may thus be maintained; or serious and fatal inflammation be set up; and concretions may be found in the urinary bladder, which will call for surgical interference for their removal.

Dr. Williams (*op. cit.*), referring to the renal irritation frequently caused in these cases (p. 130, American edition, foot-note), says: "I have in several instances found in the cortical and tubular structure of the kidney, clustered crystals of lithic acid, which, under the microscope, exhibited such sharp angles and dagger-shaped projections, as would afford an easy explanation of the pain, inflammation, and hemorrhage, often attendant on an attack of renal gravel, even when none is obvious in the urine." The same writer reminds us (p. 187), of "the proximity in composition between lithic acid and urea," and that it is probable, according to Liebig, that the former may be converted into the latter. He likewise calls attention to the fact that both gout and rheumatism may give rise to fluxes and catarrhal affections, as oliguria does. Rheumatism and Bright's disease are also, often, nearly related. In respect to this fact, Dr. Garrod announced the following opinions and conclusions, in his paper already quoted (*Med.-Chir. Transactions*, vol. xxxi.), "Blood from patients suffering from Bright's disease and albuminuria after scarlatina, was then examined; the results of these analyses appear to show that—

"1st. Uric acid is always present in the blood in albuminuria. The quantity, however, greatly varies: when the functions of the kidneys are much impaired, it exists in quantities almost as great as in gout; in other cases its amount is small, but it usually exceeds that found in ordinary blood.

"2d. Urea always exists in large quantities in this blood (a fact which has been long since proved), and no relation is found between the amounts of urea and uric acid.

"3d. The kidneys are always deficient in their power of throwing off urea; but with regard to the uric acid, their excreting function may be impaired or not." (*Loc. cit.*)

We thus see what serious disturbances may arise in the system by a perverted condition of secretion, arrest of excretion, and attempts at vicarious elimination of a product which must, in order to the preservation of health, be discharged from the body. The diseases thus produced come clearly under the head of *disordered vital chemistry.* Thus, when the above vital functions are weakened, or totally disabled, there must be not only general disturbance, but, after a time, some special manifestation of disease, and the results of vitiated secretion, decomposition, and over-worked and irritated organs.

The *contracted kidney*—called, by the late Dr. Todd, "the gouty kidney"[1]—and which, as the term he has given it implies, he considered due to the effects, at once irritant, inflammatory, and destructive, arising from gout, we may mention as a result primarily dependent on disorder of the "uric-acid-excreting" function. Dr. Todd has given satisfactory proof (*Clinical Lectures on Certain Diseases of the Urinary Organs*) from cases of patients, of the existence of this form of renal disease. He mentions, also, that many might be inclined to refer it to advanced Bright's disease; but he has signalized it in those who he believes never had that affection. The kidney is shrunken, "and its structure condensed—a condition which, while it may also occur in other states of the system, is peculiarly apt to be developed in the inveterate gouty diathesis." (*Loc. cit.*, p. 313.) In one case, necroscopy disclosed hypertrophy of the heart with dilatation—partly due to incrustations on the mitral valve—hardened, condensed, and somewhat contracted liver—the Glissonian capsule being hardened and thickened. The morbid alteration in the latter organ is explained partly by the intemperate habits to which the patient was addicted, "but partly likewise by the share which the liver had in the elimination of the morbid poison of gout." This latter result, which is not infrequent, it

---

[1] A term objected to by Dr. Barclay (*op. sup. cit.*); who, however, very distinctly recognizes the close connection between gout and renal disease. Valleix, writing in 1853, would not admit that special form of nephritis which is referred to the gouty diathesis.

is important to notice, as one of the secondary effects of the diverted and retained urinary element. The kidneys in this patient were very much contracted, being hardly one-third of their natural size; they were granular and shrivelled upon their surfaces; the investing capsule was apparently denser and whiter than usual, and was easily detached from the glandular surface. Diminution of the cortical renal substance was the source of the decrease in the size of the organ; two-thirds being estimated to have disappeared. The granulations were noticed likewise upon the cut surface. Dilated and scantily-lined *tubuli uriniferi* were observed on microscopic inspection; from some of them all the epithelial lining had vanished; others were collapsed, folded, and crumpled up; and looked like "fascicles of fibrous tissue." A few fatty epithelial cells were detected in certain *tubuli;* and others of the latter were seemingly healthy, especially those in the pyramids of Malpighi. Dr. George Johnson[1] has described, very minutely, this condition of the kidney, and particularly notes the changes which supervene in the vascular system of the gland. In another instance, Dr. Todd has recorded the discovery, in an inflamed "gouty kidney," of "*opaque streaks of deposit of lithate of soda*" in some of the renal cones. These streaks took the direction of the tubes, and certain of the latter were probably occupied by them.[2] We may well exclaim with Dr. Todd, as we reflect upon these vital manifestations and post-mortem revelations—

"How strikingly do these consequences of the long continuance of the malady comport with the humoral view of the pathology of this disease! Not only are those parts which the morbid matter of gout is most prone to affect, materially damaged, but likewise the emunctories through which the poison would make its escape out of the system—the liver and kidneys: these organs have become poisoned by the morbid matters which have escaped, or tried to make their escape from the system through them; and, therefore, it is natural to expect a considerable change in their nutrition."

It is notorious, however, that in most, if not all, of the *metastases* of gout, there are no traces of morbid action upon the organs affected by the repercussion of the disease. The reason of this doubtless is, that when a fatal result occurs in this manner, the morbid action lasts too short a time, notwithstanding its violence, to leave structural traces. Probably a true *spasmodic* action often destroys the patient, in the thoracic and abdominal varieties of retrocedent gout. It is, however, very likely that in many cases of sudden death from these causes, the vital organs have been previously weakened by disease of some sort; and in many cases chronic gout has inflicted a certain amount of injury, for, as we have already seen, the

[1] On Diseases of the Kidney.
[2] This appearance is also noted by Dr. Garrod, who saw "streaks of white matter at the apex of each pyramid and running up in the direction of the tubuli. Kidneys pale and contracted; cortical portion shrivelled." (p. 199.)

continued and recurring malady *does* leave its indications, very decidedly, behind it.

After gout in the *stomach*, Dr. Todd signalizes a "dilated and flaccid state of the organ" as existing, and this is the more marked in proportion as the attacks have been frequent.[1]

In addition to the abundant and often astonishingly copious deposits of the urate of soda into and around the joints, that salt has been found to cover and even to penetrate into the texture of the cartilage investing the affected joints (Watson), and to insinuate itself into the substance of tendons and ligaments (Dr. Wm. Budd). A curious locality is at other times chosen by it—viz., under the skin covering the cartilages of the ears. It has been remarked also over the cartilages of the alæ nasi (Todd).[2] Pus sometimes forms around the variously located depots of urate of soda, and occasionally the discharge of this liquid is quite abundant. Generally speaking, also, the joints of the hands and arms exhibit more plentiful deposits of the urate than those of the lower extremities, except in some forms of acute sthenic gout, when the reverse may be observed. The

[1] Op. cit.

[2] Dr. Garrod (op. sup. cit.) states that "sometimes small nodules of urate of soda are found upon the eyelids, especially the lower, now and then in the integuments of the face." He also refers to the great care which is necessary in ascertaining the precise nature of the "white-looking deposits" often occurring about the eyelids and face. Dr. H. Barker saw them on the nose. Dr. Garrod has "observed a true gouty deposit as large as a split pea, apparently attached to the fibrous structure of the corpus cavernosum penis." (p. 86.)

In relation to gouty affections of the eye and ear, Dr. Garrod further remarks : "A form of ophthalmia connected with gout has long been noticed." He adds that it is liable to be confounded with rheumatism, when that is directed to the ocular region. Cases of ophthalmia evidently connected with the gouty diathesis have been observed by Dr. G.; these were instances of conjunctivitis and sclerotitis. "Gouty iritis is also said to occur." The nodules found upon the cartilages of the ear have been mentioned. "Deposits are not unfrequently found upon the drum of the ear, and have been especially pointed out by Mr. Toynbee, but I have failed to discover a trace of uric acid in several examinations of them. Whether they are ever connected with gouty inflammation, is at present a matter of uncertainty ; they should especially be sought for in gouty subjects in whom the joints are much affected with chalk-stones, for if not found in such cases, it is not probable that they would occur in others." (pp. 515, 516, op. cit.)

Accomplished aurists have pointedly alluded to the connection between gouty and rheumatic affections and deafness. Mr. William Harvey (The Ear in Health and Disease, London, 1856) specially considers this subject; and we have heard him insist upon the frequency of the connection, whilst observing his aural practice.

In his elaborate work on The Diseases of the Ear, just published (London, 1860), Mr. Toynbee refers to the subject, and furnishes interesting illustrations of the reality of the connection, as observed by himself. He says : "The poison of *gout* may also give rise to deafness and other peculiar symptoms in the head." (See the work cited, page 362, English edition.)

interference with the motion of the various joints thus affected is so evident and so familiar to practitioners, and indeed so well known to nearly every one, that we hardly need do more than allude to it as a "morbid effect" of the disease. The fact, moreover, has already been made sufficiently prominent.

RHEUMATISM.—Nearly related to gout is the disease known by the term *rheumatism;* an affection which, like its *congener,* may be *acute* or *chronic,* and is characterized by very similar constitutional symptoms and by many local phenomena of like nature, while yet there are several striking points of difference. Thus, we observe indications of irritative action and of inflammatory states of variable severity ; fever, anxiety, restlessness, chills, with full, quick and sharp pulse; gastric and intestinal disturbances; thirst, white tongue, foul breath, and acid eructations. Locally, there is heat, pain, and tense, or sometimes doughy, shining redness, with swelling around certain joints, and often affecting neighbouring tissues, as the muscles, tendons, &c. By preference, the larger joints are attacked in contradistinction to gout, which, as we have seen, fixes upon the smaller. There is also the remarkable and very distinctive symptom, or rather local manifestation, of *acid perspiration,* often very profuse and offensive. The latter is apparently the corresponding fact, so to speak, to the local appearance of urate of soda in the tophaceous deposits in gout ; for, in true rheumatism, nothing of the sort occurs; in *rheumatic gout,* as will be seen, there is an analogous deposit.

From what we have already indicated, it is not difficult to see a great resemblance between gout and rheumatism ; they have indeed been denominated *"first cousins."* The first and most important question for us, concerning rheumatism, is—and it is an inquiry of great practical interest—*does it depend upon excess of uric acid in the blood, and its retention therein ?*

The latest views of the pathology of the disease seem to give a directly *affirmative* answer to this question ; and, of course, under this aspect of the subject, it devolves upon us to enumerate the affection and its results amongst those "morbid effects" of retention in the blood of the urinary "element" we are now examining. It should, however, be distinctly premised, that many researches and much close study and observation are yet required in order to enable us positively to declare uric acid or any of its products to be the *materies morbi* in rheumatism. Observers, however, at the present day, at least begin to speak with more confidence and decision upon this point. In the first place, we find it more than ever common for systematic writers on medicine to consider gout and rheumatism under one head ; often in one chapter. Next, if we examine the language employed in discussing the yet unsettled question as to the *essential cause* of the

disease, we shall observe a tone of much greater decision than in the works
published only a comparatively short time since; while the matter is yet
left open to doubt—most authorities not fully compromising themselves.
We will refer to a few of these opinions.   Dr. Barlow (*Manual of the
Practice of Medicine*, 1856) says :—

   "The external cause [of rheumatism] is generally exposure to cold, especially
when producing repressed perspiration.   Its internal or essential cause seems to
be an abnormal condition of the blood, which contains always an excess of
fibrin and of uric acid : the latter is probably the *materies morbi* or peccant
matter." (p. 130, *op. cit.*)

   Dr. Bennett (*Clinical Lectures on the Principles and Practice of
Medicine*, American edition, 1858) begins his remarks upon the General
Pathology of Rheumatism and Gout, as follows :—

   "The present theory with regard to these affections is, that they are both
connected with an increase of lithic acid in the blood.   In rheumatism, this is
dependent on excess of the secondary, and in gout on excess of the primary
digestion." (*Op. cit.*, p. 909.)

   The latter author also refers to the fact that a considerable amount of
lactic acid is excreted from the skin, as was taught by Prout, and mentioned by Todd, Watson, and others of note, who have carefully studied
the subject.   Hitherto, the tendency seems rather to have been to ascribe
the morbid phenomena of rheumatism to an excess of the last named acid
in the system.   Dr. Prout, in the fifth edition of his celebrated work on
Stomach and Renal Diseases, refers pointedly to this explanation, as being
his own belief.   While mentioning in a foot-note (p. 549) the opinions of
Dr. Garrod as to the relations of uric acid to gout and rheumatism, he
says : "In various parts of this work I have spoken of lithic acid as being
characteristic of gout, and the lactic acid of rheumatism.   At any rate, I
agree with Dr. G. that lithic acid has little to do with pure rheumatism ;
though it is often present in what is called rheumatic gout."   In the latter
affection—which is understood to be that variety of gout which attacks all
the joints, indiscriminately, and for which Dr. Todd prefers the name of
"*general gout*"—it is natural to believe that the chief morbific agency is
uric acid, or a salt from it.   It has, however, as we previously stated, until
somewhat recently, been the persuasion of the majority of observers that
lactic acid held that relation to rheumatism, which it seemed plausible to
assign to uric acid in reference to gout.   In his last edition (1857), Dr.
Watson says, in the course of a remarkably clear and accurate *resumé* of
the distinctions existing between gout and rheumatism (*op. cit.*, vol. ii.
p. 761, English edition) : "Gout is often, rheumatism is never, associated
with chalk-stones ; and conformably with this distinction, Dr. Garrod has

taught us that uric acid in excess is present in the blood of the gouty, and not present in that of rheumatic patients."[t]

The manner in which Dr. Todd (*op. sup. cit.*) refers to the method pointed out by Dr. Garrod for detecting uric acid in the serum of the blood, in gouty cases, seems decidedly indicative of his own views as respects the existence of uric acid in excess in the blood in gout and not in rheumatism ; and in this connection, we are glad of an opportunity to give the method of Dr. Garrod, above referred to.   Dr. Todd says :—

" Dr. Garrod has made out a positive physical character of gout, which may be regarded as surely diagnostic of that disease from rheumatism. It consists in the discovery of uric acid in the blood-serum or the blister-serum. And his process is ingenious, and so simple that any one may use the test, however little accustomed to chemical manipulation. A little serum is put into a watch-glass, and to it are added five or ten drops of acetic acid. In this acidulated serum a small skein of worsted is laid, and the watch-glass is set aside under cover to protect it from dust. After a few hours, the crystals of lithic acid, if it exist, will be found adhering to the threads."[2] (*Op. cit.*, pp. 408-9.)

Dr. Henry William Fuller, of London, Assistant Physician to St. George's Hospital, in his very scientific and able treatise on *Rheumatism, Rheumatic Gout, and Sciatica*, published in 1852, takes the ground that *lactic acid* is the *materies morbi* in rheumatism. He thus falls in with the opinions of Drs. Prout, Todd, Williams, and others, who so believed, although not positively asserting the fact as indisputable.   Dr. Fuller considers the cause of the disease to be a poison in the blood, and which is generated in the system as the product of a peculiar form of mal-assimilation—of vicious metamorphic action.   This poison it is "which excites the fever, and produces all the pains and local inflammations which are often found associated in an attack of rheumatism." (*Op. cit.*, p. 28.) He then points out the fact that although the fever may be increased by the occurrence of the local inflammations, "it is essentially independent of them," and often is well developed before they begin.   If the virus is in small quantity in the blood, then only slight wandering pains are produced;

[1] Speaking of Dr. Prout's belief in regard to the *lithic* and *lactic acids*, and their relations to gout and rheumatism, Dr. Watson relates certain experiments, by Dr. Richardson, upon animals. Lactic acid introduced into the peritoneal cavity of a healthy cat, produced irregular cardiac action in two hours. The animal was found dead the next morning, and no peritoneal inflammation was discovered, but "marked endocarditis of the left chambers of the heart. The mitral valve was inflamed and thickened, and covered on its free borders with firm fibrinous deposits. The whole inner surface of the ventricle was highly vascular." Similar results were observed in a dog experimented upon in like manner. No textural alterations were found in the joints.

[2] In Dr. Garrod's late work (*Gout and Rheumatic Gout*) we find this process much more elaborately described. See pages 110-113, *op. cit.*

if in large quantities, and increasing, the effects are usually proportionate. Some constitutions, of course, manifest more resistance than others.

We have previously cited the authority of Dr. Garrod in reference to the morbific agency in gout and rheumatism;[1] and in his late work, we find him reiterating his views, and pronouncing the urate of soda "*pathogno-monic*" of gout. Some of this author's conclusions thus lately arrived at, or rather confirmed—for such, we believe, has always been his doctrine—are so positive and so much to the point for our present purpose, that we quote them. His analyses show that "healthy blood contains the merest trace of uric acid or urea, so small as to be in general undiscoverable, except by the most minute and searching chemical examination, and not always then.

"That, in gout, the blood is invariably rich in uric acid, which exists in the state of urate of soda, and can be separated from it, either in the form of the crystalline salt in acicular needles, or as rhombic crystals of uric acid.

"That, in acute rheumatism, the blood is free from uric acid, or, at least, contains no more than in health.

"That the perspiration seldom contains uric acid; but that, in gout, oxalate of lime may be crystallized from it, as also from the blood."[2]

In the above expressions of opinion, founded on careful analysis, we find Dr. Garrod distinctly declaring that the blood is free from uric acid, or only contains a normal amount thereof; he therefore does not consider that acid the cause of rheumatism.

Such are some of the opinions to which we referred ; and while, in view of the existing difference in the decisions of equally eminent men upon this point, we cannot look upon the pathology of rheumatism as by any means definitively settled, it seemingly devolves upon us to consider the disease as a condition referable to excess of uric acid in the system, in deference at least to the opinion of many of the latest observers, whose position, judgment, and opportunities for the accumulation and weighing of evidence are such as to entitle their announcements to the greatest respect. In this regard, however, as much can be said for those who hold the opposite views, or who do not fully compromise themselves; but we shall be quite safe, at least, in discussing the question according to the plan above announced. Our own conviction, hitherto, has always been wholly in favour of the *lactic acid* theory ; and, within a day or two, having put the question to a highly-cultivated and well-informed medical friend—"What do you consider the *materies morbi* in rheumatism to be ?" his reply was, after a

[1] "The Nature and Treatment of Gout and Rheumatic Gout."

[2] In giving these opinions of Dr. Garrod, we followed the statement furnished in the *Lancet* of December 24th, 1859, not then having seen Dr. G.'s work. The account is by the Reviewer, condensed from the book itself. We have lately, however, had the volume at our disposal, and can testify that it is a most satisfactory, complete, and erudite treatise.

few moments' consideration, "Some product of *lactic acid*—some of the lactates."

Whatever the fact may be, it is clear that the fault of *deficient excretion* is equally active in rheumatism as in gout; the former is as distinctly a blood-disease as the latter.[1] The deranged excretory function ("Uric-acid-excreting"), however, chooses a different field for the manifestation of its disorder or impairment; the disarranged balance between the excess of acid and the excreting power being as evident in one as in the other. Dr. Bennett (*op. cit.*) writes very clearly and comprehensively on this point. He remarks: "In both diseases there is an undue balance between the excess of lithic acid and the power of excretion—in rheumatism by the skin, and in gout by the kidney. This pathology serves to explain the similitudes and differences existing between the two affections. In both there is a certain constitutional state dependent on deranged digestion, during which, exciting causes occasion local effects."

He then signalizes the fact that in rheumatism the *exciting* causes are those of a depressing nature, and are usually exerted upon the poorer classes. The chief provocatives of rheumatism, as is well known, being cold and wet, bad and insufficient food, and hard labour. As we have already set forth, precisely the opposite immediate causative elements are active in gout, *i. e.*, luxurious and intemperate diet, indolence and self-indulgence of all kinds. Dr. Fuller's idea—and which is, doubtless, that of every reflecting and intelligent practitioner—is, that the morbid material acting in the blood, is often, if not nearly always, the *sole* cause of rheumatism—secondary or exciting causes not coming at all into play, or proving ineffective, if brought to bear on the system. They are only promotive agencies, not causative elements.

It will not be expected of us, we conclude, specially to detail the course of rheumatic fever, or of chronic rheumatism. Indeed, with the present undetermined state of the pathology of the disease, we might perhaps have deemed it justifiable to withdraw the disease from the category of the "morbid effects" of retention of an element of the urinary secretion in the blood. It could hardly, however, be other than an omission of some moment, had we not carefully examined the present belief on the subject; and, having done this with considerable research, we place in due order the names of all the authors of most eminence in regard to the subject, within the last few years, whose works we have been able to consult. Their opinions are expressed with more or less positiveness; some only implying their belief, others boldly asserting it.

[1] "In truth, acute rheumatism is a blood-disease. The circulating blood carries with it a poisonous material, which, by virtue of some mutual or elective affinity, falls upon the fibrous tissues in particular, visiting and quitting them with a variableness that resembles caprice, but is ruled, no doubt, by definite laws, to us, as yet, unknown." (Watson, *Practice of Medicine*, vol. ii. pp. 738-9.)

What is the Materies Morbi in Rheumatism?

LACTIC ACID.				URIC ACID.			
Dr. Prout	.	.	. (1848.)	Dr. Barlow	.	.	. (1856.)
" Fuller	.	.	. (1852.)	" J. H. Bennett	.	. (1858.)
" Todd	.	.	. (1857.)	" Copland	.	.	. (1858.)
" Watson	.	.	. (1857.)	" Thudichum	.	.	. (1858.) [1]

With respect to Drs. Thudichum and Golding Bird—the latter of whom might have been ranged with the former in the above table—their opinions are inferred from expressions in their works, such as we quote in a footnote.[2]

It may here be added that Dr. Prout intimates it to be a very supposable and plausible doctrine that the phenomena of rheumatic gout—which, by the way, both he and others pronounce very difficult to relieve—may be explained by the fact of the concurrent action of both lactic and lithic acid in the system. This form or combination of disease requires to be described more at length, as being less familiar than rheumatism, and presenting more novel points for inquiry.

RHEUMATIC GOUT.—We have already referred to this affection, which presents a singular combination of the characteristics of rheumatism and gout. Without resembling either, as a whole, it partakes, seemingly, of the nature of both. Unlike gout in general, it attacks the weakly individual as well as the strong, and quite as readily; or else those who are robust, when either physically or mentally depressed. In this malady the analogy of the two diseases is most apparent; or rather, their dependence upon a

[1] Dr. Garrod might, by inference, be placed in the left-hand column; since, although he does not, so far as we have been able to ascertain, say that rheumatism depends on the presence of an excess of *lactic* acid in the blood, he demonstrates the absence of *uric* acid in the cases he has examined, and evidently does not consider it, in any wise, dependent thereon. (See his late work on *The Nature and Treatment of Gout and Rheumatic Gout.*)

[2] "In the two allied affections, gout and rheumatism, exclusive of the many neuralgic diseases popularly referred to the latter, a remarkable tendency to the formation of an excess of uric acid, both pure and combined, occurs." G. Bird, *Urinary Deposits*, English edition, by E. Lloyd Birkett, M. D., 1857, p. 150.

"As a question clearly put is half the answer, we may be permitted here to consider what proximate conditions of the system a rise or fall in the quantity of uric acid beyond the normal limits is likely to indicate. A deficiency may be due to a diminished production in the system, as in anæmia, or to retention, as in certain stages of gout and rheumatism. It is at least questionable whether the retention is *always* due to diseased action of the kidney. Any disease, however, which interferes with the secreting power of the kidney by changing its structure, such as Bright's disease, is certain to cause retention of uric acid in the blood, in proportion to the retention of the other constituents of urine. (Scarlatina seems to make an exception. * *.") Thudichum, *op. cit.*, p. 95.

similar cause is perhaps most clearly seen. We observe, however, that Dr. Garrod, in his work just published,[1] does not incline to the idea that this affection is actually a combination of gout and rheumatism. He prefers, moreover, to employ the term "*Rheumatoid Arthritis*," and remarks: " If we agree to name a disease simply from its external characters, then I admit that the term rheumatic gout is not inappropriate ; but if we advance further, and have regard to more intimate pathology, then I deny the propriety of the name : acting upon the former principle, we should be equally justified in calling some cases of scarlatina or measles by the compound term of rubeolo-scarlatina, and we know that these diseases were not separated two centuries ago * *." " Although unwilling to add to the number of names, I cannot help expressing a desire that one might be found for this disease, not implying any necessary connection between it and either gout or rheumatism. Perhaps *Rheumatoid Arthritis* would answer the object, by which term I should wish to imply an inflammatory affection of the joints, not unlike rheumatism in some of its characters, but differing materially from it." (*Op. cit.*, pp. 533–4.) Dr. Garrod also refers to the fact that but few authors recognize any existing combination of gout and rheumatism, many even strongly oppose such a doctrine. He mentions Boerhaave, Van Swieten, Cullen, Heberden, and Watson, as not alluding to such a connection or fusion, and quotes Sir C. Scudamore, as saying "that the textures which have been long affected with gout, become so much weakened as to be very susceptible to vicissitudes of temperature ; and in this way the general disorder may partake of rheumatism. It was only thus that he could attach any propriety to the very common expression *rheumatic gout.* It would appear that the term is often made use of, but seldom attempted to be defined with precision." (*Op. cit.*, p. 527.)

The affection has been known by various names. Thus Dr. Haygarth styled it " *Nodosity of the Joints ;*" Cruveilhier, " *Usure des Cartilages Articulaires ;*" Dr. Adams, of Dublin, " *Chronic Rheumatic Arthritis.*" Dr. Garrod, who mentions the above designations, also says : " It has sometimes received a name dependent on its situation ; for example, rheumatic gout, when the wrists, hands, and feet are affected ; chronic rheumatism, when in the shoulder, elbow, or knee, either singly or simultaneously ; and *morbus coxæ senilis,* when located in the hip. Occasionally it assumes an acute, or rather subacute, character, but more generally it is chronic throughout." (p. 533.)

Neither *sex* nor *condition* in life seems to have any influence in procuring immunity from this disease. According to Dr. Adams, the hip-joint is most often affected in males ; in females, the wrists and hands. He found

---

[1] The Nature and Treatment of Gout and Rheumatic Gout. By Alfred Baring Garrod, M. D., F. R. S., Fellow of the Royal College of Physicians, Physician to University College Hospital, etc. etc. ; London, Walton and Maberly, 1859.

it more prevalent amongst the poor and labouring classes; Sir Benjamin
Brodie considers it most common in the higher classes. This diversity of
opinion, Dr. Garrod explains by referring it to the "difference in the class
of cases more prominently brought under each surgeon's notice." (*Op. cit.*,
pp. 534-5.)

The *main characteristics* of the disease may be grouped as follows:
Pain, of a migratory nature, and which is often very severe and obstinate.
It is sometimes aggravated at night, and also by heat. (Garrod.) Motion
is extremely painful to an affected joint, especially after prolonged rest.
There is notable frequency of recurrence, and predilection for the smaller
joints. The joints attacked become, after a time, swollen; fluid is effused
in considerable quantities into the cavities of the joints, and fluctuation is
often perceptible. The ratio of helplessness is, of course, in relation with
the number of joints crippled; generally, none of the articulations escape,
when once any of them have been invaded. There is usually but little
accompanying febrile excitement and constitutional sympathy; although
this depends, very distinctly, upon the number of joints attacked; that is,
upon the extent, or degree of universality of the disease. (*Auct. cit. et al.*)

Complications.—Cerebritis, pleuritis, inflammation of the eye. The latter
is comparatively rare; all the ocular tissues may be affected; generally, however, if the patient have treatment, the choroid coat and the iris escape.
Exhausted and shattered constitutions are those most obnoxious to ocular
inflammation in connection with rheumatic gout. (Fuller.)

Results.—Thickening and permanent enlargement of the structures
forming, and (less frequently) of those surrounding the joints; occasionally,
there is cuticular desquamation, as in gouty cases. In the chronic form,
which, as we have stated, is most commonly observed, the articular cartilages are affected, the ligaments about the joints are stretched, and the ends
of the bones are irregularly enlarged. The synovia subsequently becomes
absorbed, and the capsular membrane is left much thickened. The *liga-
mentum teres* of the hip, and the tendon of the biceps are sometimes destroyed, and even completely removed. Not only the articular, but the
inter-articular cartilages are absorbed; this is observed in the knee-joint,
the wrist and the lower jaw. (Garrod.) When the disease has been very
prolonged and severe, the synovial membrane becomes not only thickened,
but droops into the articular cavity; or, as Dr. Fuller points out, "a
dense, ligamentous substance, resulting probably from some peculiar alteration in the synovial membrane, is seen interposed between the articulating
surfaces; or small irregularly shaped cartilaginous bodies are found, either
loose within the joint, or attached to it by pedicles formed of thickened
synovial membrane." Sometimes these excrescences are bony; and others
of a vascular nature likewise exist. (Garrod, *op. cit.*) The opposite surfaces of the bones forming the joints, being denuded by chronic wasting of
the cartilages, and rendered smooth by attrition upon each other, are

found to be white, glistening, and ivory-like in appearance.[1] The latter condition, we conclude, is that observed in what is termed "dry chronic arthritis"—*Arthrite chronique séche* of the French writers. We have had the opportunity of witnessing this alteration in several instances; the state is that known as *eburnation*. In specimens which we have seen, the change of texture was chiefly marked in the track of the *greatest attrition*, in the affected joint; a fact noticed particularly by Dr. Garrod. (*Op. cit.*)

"The denuded surfaces become partly worn away, and a smooth enamel is formed by the mutual action of the bones on each other, and around the articular surfaces thus acted upon bony vegetations arise.

"In most joints, after the fluid has become absorbed, a crepitus is felt on movement from the rough surfaces grating against each other." (p. 540.)

There is, in certain instances, a pulverulent deposit, usually consisting of lithate of soda, but containing also, at times, lithate of potash, ammonia and lime, chloride of sodium, phosphate and carbonate of lime. This lightly incrusts either the entire surfaces of the affected joints, or only portions thereof; and it may pervade their solid structures.

Although this deposit is identical, or nearly so, with that of gout, and occurs alone in those cases of rheumatic gout most nearly resembling genuine gout, yet, says Dr. Fuller, "I cannot therefore admit the conclusion that the existence of such deposit is of itself sufficient to mark such cases as examples of true gout." (p. 331.) The reasons he assigns for this opinion are, briefly : 1. The occurrence of these cases in "thin, spare persons of temperate habits," who have had rheumatic rather than gouty symptoms; 2. The difference in form and situation from the chalk-stone deposit of true gout; 3. The disease occurs, often, in those who formerly have had pure uncombined rheumatism; and these persons sometimes exhibit the well-known external traces of that disease. This author distinctly avows his belief that rheumatic gout is a true combination of the two diseases whence it derives its name; and when, on dissection, one joint is found to present most evidence of the gouty, and another most of a rheumatic element having been at work, he explains this by supposing that more of one influence was in force during the time such a joint was affected; and the other result rests upon similar action from the opposite influence. Dr. Garrod, who, as we have already stated, does not consider the affection to be a compound of gout and rheumatism, admits, notwithstanding, its greater resemblance to rheumatism than to gout; but he believes that much harm has been done practically, by confounding it with either gout or rheumatism.[2]

[1] Fuller.
[2] We take this opportunity to refer those interested in these subjects, to the elaborate and interesting treatise of Dr. Garrod, which we have designated above. The differential diagnosis of gout and rheumatism is clearly set forth, and the work is fully illustrated by tables, plates, etc.

4

Such, then, are the principal results—if we adopt the theory received, as we have seen, by so many reliable observers, viz., that rheumatism depends on retention of uric acid in the blood—of this poisoning of the blood by the undue presence of the *materies morbi* in the system ; and both in this and the previously considered affection (gout) we have a series of symptoms indicative of the great extent to which the infection of the blood sometimes' attains ; and also ocular evidence, both during life and after death, of the power and virulence of the poison.[1]

We have not treated of the occurrence of *metastasis* of rheumatism under a separate head, because the chief danger in this light is of the *heart* being affected, and we have already pointed out the frequency of such attacks, and the necroscopic phenomena. Sufficient allusion, we believe, has also been made to the occasional transfer of the affection to other organs. The *sequelæ* of the disease have likewise been incidentally mentioned under the head of *results* and *consequent diseases*. It is, however, not supererogatory again to refer pointedly to the extreme danger which environs those who, during rheumatic fever, have also had heart disease. Such persons, even if apparently entirely recovered, *are never to be considered out of danger*, and they require, of course, due warning from their physicians, and the exercise of great discretion on their own part, in order that they may be fully aware of their insecurity, and able, so far as is possible, to guard against a return or re-awakening of their formidable malady.

We have thus endeavoured to present the chief phenomena, resulting directly and also secondarily from retention of uric acid in the blood. Whilst the manifestations of *gout* and *rheumatism* have been, of necessity, prominent, the reader will not have failed to remark the lengthy list of antecedent, concomitant, and subsequent, ailments, seemingly more or less dependent upon that morbid influence which, by common consent, is allowed to be the active promoter, and, we may say, the originator of the disease. From the slighter derangements of the general health, through all the phases of dyspeptical disorder, to the agonizing paroxysms of gout and rheumatism, their *sequelæ*, exceedingly dangerous metastases and disastrous ravages upon the external form, both as to appearance and freedom of motion ; and finally in the *post-mortem* evidences of its activity, do we not seem to recognize the presence and morbid power of that

---

[1] It should be stated that the testes, as well as the skin, periosteum and aponeuroses are occasionally affected, either concomitantly and acutely, or subsequently and in a more chronic manner. Dr. Fuller and others have enlarged upon the predilection of rheumatism for "the white fibrous tissue." The former author refers to the mention by Dr. Watson, of rheumatism of the articulation of the jaw, and also of that of the membranes of the spinal cord by Dr. Copland and others. (*Op. cit.*, p. 46.) We need hardly refer to the fact of the frequent association of rheumatismal attacks—in predisposed subjects—with gonorrhœa.

"element of the urinary secretion" we have been considering, prevented as it is in such cases from obtaining an exit through its natural excretory channels?

CREATINE AND CREATININE.—*Formulæ:* Creatine, $C_8H_9N_3O_4+2$ Aq. —Creatinine, $C_8H_7N_3O_2$.—[1]

Creatine.—This substance, discovered by Chevreul, in 1835, is crystallizable, and is derived from the juice of flesh. It is stated to be "present in the blood and urine of man and of all animals hitherto examined." (Thudichum, *op. cit.*) Both this substance and creatinine, in the form of chloride-of-zinc salt, were found in the urine by Heintz and Pettenkofer, simultaneously, in 1844; but these chemists did not then recognize their identity with the above-mentioned product of the juice of flesh. Liebig, in 1847, demonstrated both creatine and creatinine to be "constant ingredients of the juices of the flesh of nearly all the classes of vertebrate animals, and of the urine of man." (*Idem.*)

Although now a recognized element of the urine, creatine is contained therein in an exceedingly minute proportion. It is always present in the blood, which is a fact important to note in respect to our present inquiry; for, if naturally existing in the vital fluid as a product of a chemical muscular change or waste, it is doubtless innocuous so long as it is finally

[1] Thudichum.

Chemical Composition of Creatine and Creatinine.

Creatine:	8 equivalents of carbon	.	.	.	48	36.64
	3 " " nitrogen	.	.	.	42	32.06
	9 " " hydrogen	.	.	.	9	6.87
	4 " " oxygen	.	.	.	32	24.43

Atomic weight of dry creatine   131   100.00

To procure the formula for "crystallized creatine,"

Take, 1 atom of dry creatine	.	.	.	131	87.92
and 2 atoms of water	.	.	.	18	12.08

149   100.00

$C_8N_3H_9O_4+2HO=131+18=149.$
$C_8H_{11}N_3O_6.$

Thudichum.
G. Bird.
G. Johnson.

Creatinine:	8 equivalents of carbon	.	.	.	48	42.48
	7 " " hydrogen	.	.	.	7	6.19
	3 " " nitrogen	.	.	.	42	37.17
	2 " " oxygen	.	.	.	16	14.16

Atomic weight of creatinine   113   100.00

$C_8N_3H_7O_2=113.$
$C_9H_7N_3O_2.$

Thudichum.
G. Bird.
G. Johnson.

The formula expresses "crystallized" creatinine.

discharged by the kidneys, its proper emunctories.[1] And unless we suppose extensive renal disease and destruction of tissue or serious obstruction to the excretion of urine, it is not possible that undue accumulation of this substance in the blood should take place. Authors tell us that it is to be considered as undoubtedly excrementitious (Golding Bird, Thudichum); therefore the blood must be depurated from it as it is formed and thrown in upon the circulation, or doubtless the phenomena of systemic disturbance, such as would arise from contaminated blood, would occur. And with this view, we should expect to observe somewhat analogous morbid phenomena to those produced by uræmic poisoning—bearing in mind the close relation to urea which is held by creatine, as has already been mentioned, on the authority of Dr. Thudichum.[2] This supposition, which naturally occurred to us in view of the *excrementitious* nature of the substance in question, seems to find confirmation in the following remarks of the last-named eminent observer. " In disease, the quantity of creatine, together with that of creatinine, might serve to indicate the intensity of any spasmodic or convulsive action. The question as to its quantity in tetanic and epileptic disease is one of high interest. Cases of paralysis agitans, in which the spasmodic action ceases with sleep, may perhaps afford good opportunities for demonstrating the influence of rest and motion ; though the different nutrition in the muscle may, perhaps, vary

[1] " Creatine is present in the blood, by which it makes its way to the kidneys. It occurs in the urine as a regular ingredient, though present in small quantities only. It is partly transformed into creatinine, most probably somewhere between the muscle and the urinary residue out of which the zinc salt crystallizes. For in the muscle, creatine has by far the preponderance over creatinine ; in the urine, creatinine over creatine. Creatine is, therefore, truly excrementitious ; its relation to urea proves this beyond doubt. Its exclusive occurrence in the muscles shows the seat of its formation ; it is, with other matters, a product of the chemical changes in the muscles." (Thudichum, *op. cit.*, p. 120.)

[2] In a uræmic case, Hoppe extracted five times the normal amount of creatine from the muscles of the patient. (Braun, *op. cit.*, p. 99.)

Dr. Golding Bird (*Urinary Deposits*, 1857) has some interesting remarks in reference to the modes of excretion of creatine and creatinine, and also in regard to their relation to urea. " Although we have seen that creatine and creatinine are both found in the urine, we must not conclude that they are entirely excreted in this manner. It is very probable that a considerable proportion of creatine is resolved into uric acid or urea before its final elimination. We have already seen the chemical relation of creatine to uric acid, and to urea ; its metamorphosis into the latter body, and into the peculiar substance sarcosin (which requires only the addition of the constituents of water to represent the elements of lactate of ammonia) is so readily effected, that a similar change occurring in the body is rendered very probable." (p. 109, English edition.) We may thus observe that it is very possible for a large amount of creatine and creatinine to be thrown out of the system in other ways than by the kidneys—a fortunate circumstance in renal disease, both when these and other urinary elements are concerned.

the chemical changes in some degree. These suggestions for future researches must not be taken for theories or suppositions." (*Op. cit.*) There are almost no *data*, so far as we are aware, which could enable us to set forth, with any precision, the effects of an undue accumulation of either creatine or creatinine, considered as urinary elements, in the blood. As the author last quoted intimates, "future researches" can alone enable observers to furnish satisfactory details ; and although he is careful to re- pudiate the idea of issuing theories or suppositions, there seems no great presumption in hazarding the latter, as we have done in a former para- graph ; especially in view of a dearth of facts which precludes our offering much upon the subject.

It is evident, however, that one practical deduction may be drawn from the fact that these substances are the result of muscular waste, viz., that if there is an excessive amount of them observed in the urine for a long time, this waste will also become extreme—consequently, rest and the appro- priate treatment for preventing debilitating action should be observed.[1]

Creatinine.—This "substance is found in the muscles of the vertebrate animals, and in the urine of man in larger quantities than in the muscles. It is the product of the natural or artificial decomposition of creatine * * *." (Thudichum.) Its component elements are carbon, hydrogen, nitrogen, and oxygen.

We have nothing further to present relatively to creatinine than what has already been incorporated under the head of creatine.

Hippuric Acid.—(Symbol : $\overline{\text{H}}$.   Formula : $C_{18}H_8NO_5 + HO$.)[2]

Liebig has announced this acid to be a constant ingredient of human urine ; a statement which, according to Thudichum, has lately been nega- tived by Duchek. It is stated that its existence was demonstrated in the urine of young infants by Scheele, Fourcroy, and Reynard. Benzoic acid, when ingested, becomes hippuric acid in the body. (Ure, *Med.-Chir. Transact.*, vol. xxiv. ; and Keller, *Ann. d. Chem. und Pharm.*, xliii. p. 198 ; Thudichum, *op. cit.*)

[1] See, among other references on this point, Dr. Hassall's work on "The Urine in Health and Disease."
[2] Or, $C_{18}H_9NO_6$.—*Thudichum*.

Chemical Composition of Hippuric Acid.

Carbon	60.335
Hydrogen	4.469
Nitrogen	7.821
Oxygen	22.347
Water	5.028
	100.000

Thudichum.

The *pathology* of hippuric acid is fully as undetermined as that of the two substances we have last considered. Its occurrence in the system in excess has generally been attributed and easily referable, to peculiarities in the diet. According to Dr. Golding Bird and some others, it seems to be found especially in those who live exclusively upon vegetables, milk, and certain kinds of fruit, and to be most constantly observed in persons of indolent habits. This is esteemed the reason why it is detected in stall-fed cattle, or in well-fed stallions, and not noticed, but replaced by benzoic acid, in cattle that are worked or driven. (*Thudichum et al.*) Dr. Bird ascribes its presence in the urine of nursing infants—a fact already referred to—to their "mal-assimilating the large quantity of carbon contained in the food." This author, moreover, does not consider that it interferes with the production of uric acid; but he observes that in hippuric urine there is generally a deficiency of urea. Reference is also made by him to very interesting cases by Bouchardat, Garrod, and Pettenkofer. In Bouchardat's case, the acid was observed to coexist with albumen in the urine,[1] and an absence of uric acid. The patient had lived for nine years on a milk diet, was fifty-three years old when the case was noted, had resided in the country, had good general health, was in easy circumstances, was the mother of one child, and had ceased to menstruate at the age of forty-five. Gastric and hepatic difficulties of an obscure nature led to her adoption of the milk-diet, and her health was restored. She then partook of a more mixed diet, eating some meat and vegetables; after a time she again became ill, and the chief symptoms were lassitude, dryness of the skin (perspiration having previously been profuse); vague pains in the hepatic region; jaundiced hue; the feces were black; the mouth dry, with a bad taste; there were headache and tinnitus aurium; imperfect vision; palpitation of the heart; excited pulse; anæmic murmur in the carotids; some œdema of the lower limbs; dyspnœa. The chief phenomena, however, were excessive thirst and increase in the quantity of the urine. She often drank from six to ten pints of water in a day. The patient died exhausted.

Dr. Garrod's case was observed in the person of a young man, twenty-three years of age. The first signs of constitutional disturbance were "general malaise," with an excretion of an excess of urea from the kidneys, accompanied by a deposit of the ammoniaco-magnesian phosphate. Atonic dyspepsia, with lumbar pain, succeeded. At this time, hippuric acid, in "long crystals," was detected, and these were slowly incrusted with uric acid. This lasted for several days, and a pint of urine yielded about 40 grains of hippuric acid. Uric acid and urea were observed in normal proportions. The hippuric acid soon decreased, and the urine finally

---

[1] Dr. Prout remarks that both xanthic oxide and hippuric acid "are undoubtedly of albuminous origin." (*Op. cit.*, p. 238.)

became normal. "No information as to the source of the hippuric acid could be obtained from the history of the patient. He had lived on a mixed diet, and never used any excess of vegetable food, nor had he ever taken any benzoic acid." (*Op. cit.*, p. 211.)

The case reported by Dr. Pettenkofer is regarded by Dr. Bird as perhaps the most interesting of the three. The patient was a girl thirteen years old, affected with chorea, and long subject to it, under the care of Dr. De Marcus, of the Julius Hospital, Wurzburg. There had also been "anomalous hysteric symptoms." She had for a long time lived on apples, bread, and water, refusing any other food. The urine was yellow, limpid, and faintly acid when first excreted; it soon became alkaline, and deposited crystals of the triple phosphate of magnesia. Adding hydrochloric acid to it, after moderate concentration there was "a copious formation of crystals of hippuric acid. The addition of nitric acid, by its oxidizing influence, caused the deposit of hippuric to be replaced by one of benzoic acid. In 1000 parts of urine there were—of

"Water . . .	. 959.332
Solids . . .	. 40.668
	1000.
Solids soluble in alcohol	18.451
" insoluble in alcohol	9.417
Anhydrous hippuric acid	12.800
	40.668

Fixed salts, containing much carbonate of soda, 19,599.

"The characters of the urine in this case approached those of an herbivorous animal, in the presence of hippuric acid and of carbonate of soda in the ash, as well as in the absence of uric acid.

"The hippuric acid disappeared, and the urine assumed its normal proportions on inducing the girl to return to a mixed diet." (*Op. cit.*, pp. 211, 212.)

We have given the substance of these cases, because they seem to bear so directly upon the portion of our subject now under examination. If, as may reasonably be inferred, the morbid phenomena, both the early and the later, may be ascribed to that state of the system which at last declared itself by the discovery of an excess of hippuric acid in the urine, we may logically argue that this was the morbific agent; and that the surcharging of the blood with it, and the contamination of the various organs by this vitiated current, is the most plausible explanation. At all events, there seems to have been a direct connection between the morbid condition and the excess of the acid in the system. Dr. Bird and others, as has been mentioned, ascribe its presence in the urine in great excess, to the use of a not sufficiently nitrogenized diet, or to mal-assimilation of the carbon of the food. It has been presumed possible that, through the kidneys, hippuric acid *vicariously* depurates the liver from any excess of carbon. (Bird.) Supposing this to be true, and renal disease to supervene under these circumstances, the most disastrous results would seem to be unavoidable.

There is no symptom or set of symptoms, so far as we are aware, which would indubitably indicate an excess of hippuric acid in the blood ; and the manifestations which we have enumerated, occurring as directly antecedent phenomena to the detection of such an excess, and seemingly dependent upon it, are only to be taken as *probably* the explanation thereof. It is, at least, reasonable to conclude, as we previously remarked, that when the acid appears, more or less suddenly, in excess in the urine, it *has been* equally so in the blood, and for a longer or shorter time—the period, possibly, being indicated more or less distinctly by certain morbid signs. Dr. Golding Bird closes his remarks upon the Pathology of Hippuric Acid in the following terms : "My own experience in these cases has been too limited to justify my offering any opinion on the pathological complications attending them. From what little I have observed, I feel inclined to believe that when an excess of hippuric acid exists, it may always be regarded as traceable to, or pathognomonic of, the deficient function of the liver, lungs, or skin, the great emunctories of carbon ; or to the use of food deficient in nitrogen. It hence follows, that our treatment will consist in appealing to the function at fault, and carefully regulating the diet.

"I would suggest the propriety of seeking for the presence of hippuric acid in the urine, where it is copious, of low specific gravity, but slightly acid or neutral, and occurring in persons who have a dry and inactive state of surface, with anæmia. In many pseudo-chlorotic cases in both sexes, I am inclined to believe an abnormal proportion of this acid will often be met with." (*Op. cit.*, p. 213.)

Dr. Thudichum refers to the discovery, by Lehmann, of hippuric acid in diabetic urine, whenever he had sought for it ; and also in the acid urine of fever-patients, "of which it is said to cause, in part at least, the acid reaction." Hünefeld and Duchek confirmed the experiments of Lehmann. Schlossberger found hippuric acid in the scales of ichthyosis. Whether this was only an isolated instance, "or whether it is of frequent or constant occurrence in that disease," is not stated. (*Auct. ante cit.*)

Many observations are needed before we can attain to any more precise knowledge of the pathology of hippuric acid. The subject is yet in that undeveloped state which induced Dr. Thudichum to conclude his chapter upon it in these words : "The reader will think this a very unsatisfactory chapter, and so indeed it is. We want observations, for which there is a large field open. But undoubtedly some technical difficulties will have to be overcome first, before the analysis of hippuric acid can be made with sufficient accuracy." (*Op. cit.*, p. 152.) This author has recently investigated the subject in its chemical, physiological, and pathological relations; and if he finds the stores of information in regard to it so meagre, and if

Dr. Prout was obliged to confess that he was not aware that hippuric acid, either in excess or in deficiency, is characteristic of any peculiar disease (*op. cit.*, p. 239), we surely need not shrink from avowing the poverty of our own knowledge in the matter.

CHLORINE : CHLORIDE OF SODIUM.[1]

Were we to follow Dr. Golding Bird's estimate of the "essential" elements of the urinary secretion, we might now pause in our examination of the list of ingredients which we at first enumerated as entering into its composition. Dr. Bird, after mentioning what he terms the "Organic Products," viz., urea, uric acid, creatine, creatinine, colouring and odorous principles, together with hippuric acid and lactic acid, which latter he also styles "accidental constituents," says that this "class of ingredients can alone be considered as really essential to the urine, and characteristic of it as a secretion, the kidneys being the only organs which normally eliminate these elements from the blood." The *saline* ingredients, as he remarks, "are met with in most secretions of the body, with the exception of the sulphates, which are rarely found except in the urine." The "ingredients derived from the urinary passages" (Dr. Bird's *Third Class*) are found "in all fluids passing over mucous surfaces, the phosphate of lime being derived from the mucus, of which it is a constituent."

Since, however, there are many interesting and important points connected with the consideration of the remaining constituents of the urine as given by Dr. Thudichum, whose table we have adopted in preparing this essay, we will bring them separately under consideration according to our original plan. What we have to present, however, will naturally be more general, since we have no affections to consider in this relation, which like gout and rheumatism seem to depend so entirely upon some morbific matter retained in the blood. That an excess or diminution of the chlorides and of other matters in the system, has a greater or less bearing and significance in certain diseases, seems to be proved in many instances, and in others rendered extremely probable. It will be our object in the remainder of this essay to illustrate these positions so far as we may be able ; bearing in mind that our main purpose is to indicate the results determined by the undue presence of these elementary substances in the circulation.

With regard to the mere presence of *chloride of sodium*—which substance we shall make the foundation of our remarks—in the blood, it is well known that there is always a greater or less amount taken into the system with the food. There is, indeed, a strong natural appetite for salt, both in men and animals; which, however, varies remarkably in different

[1] Chemical Composition of Chloride of Sodium :—

Formula : NaCl. $\left\{\begin{array}{l}\text{Sodium . . . 23.3}\\\text{Chlorine . . 35.5}\end{array}\right\}$ 58.8=1 equiv. chloride of sodium.

individuals. From the fact of the almost universal desire for it which exists, we cannot suppose that it is, *per se*, ever noxious, unless it be ingested in enormous and unnatural quantities; or unless, through other influences, the proper balance of its proportions in the blood be permanently or for a long time disturbed.

The valuable experiments of Barral,[1] Regnault, and Reiset, alluded to by Dr. Bird, Dr. Thudichum, and others, led to the conclusion that the elimination of the nitrogenized elements of the urine was facilitated by the action of the chloride of sodium. Dr. Thudichum thinks that if equally careful experiments were again conducted "by the more accurate methods" now at our command, we should acquire very important information in respect to the "causes and influences" which determine and modify the amount of chlorine thrown off by the kidneys; and this especially if the feces and other *excreta* were taken into the calculation. Barral ascertained by his experiments, both the whole quantity of chlorine taken with the food, and also that of chlorine and urea excreted. The action of chloride of sodium is certainly salutary also in another way; and which is particularly pointed out by Dr. Thudichum, viz., by its causing thirst, and consequently inducing the ingestion of an increased quantity of water, the diuretic influence of which, by producing a more copious urinary flow, "carries away not only the salt, but also organic ingredients in solution." (*Op. cit.*, p. 165.)

The Relation of Chlorine to Pneumonia and other Acute Diseases.— The very striking fact of the rapid diminution, and occasional temporary disappearance of the *chlorides* in the urine of pneumonic patients—first pointed out by Simon and Redtenbacher, and subsequently sedulously tested by several observers, amongst whom Dr. J. H. Bennett may be named as having supplied us with a large amount of clinical observation— naturally arrested the attention of pathological chemists and medical practitioners. It was at first supposed that the above-mentioned diminution was distinctive of the pneumonic inflammation; but subsequent researches seem to preclude this idea. Dr. Thudichum has lately announced the following proposition relative to this point : "The result of many observations of Vogel and others, last, of myself, then, is that *in all acute febrile diseases the amount of chlorine discharged in the urine sinks rapidly to a minimum, say one hundredth part of the quantity normal to the individual, until at last, in certain cases, it disappears entirely for a short time. When the diseased action is abating, the amount of the chlorides rises during convalescence, sometimes above the normal average.*" (*Op. cit.*, pp. 165-6.)

While, therefore, this diminution in and temporary disappearance of the

---

[1] Barral, S. A., "*Statique chimique des Animaux, appliquée specialement à la question du sel*," Paris, 1850. (Thudichum.)

chlorine is not found to be solely characteristic of pneumonia, the cases of that disease in which it has been observed afford very striking illustrations of the fact.   For detailed accounts of these, the recent work of Dr. Bennett, already referred to, may be consulted with great advantage.   Dr. Beale, of London, has also given us much reliable information upon this important topic.   (See *Medico-Chirurgical Transactions*, vol. xxxv.)   The rule in pneumonia seems to be that the diminution of the chlorides indicates the progressive stage of the disease; at its height, the chlorides may wholly disappear; their reappearance is a sign of improvement, with cessation of the inflammatory action, and occurrence of the *crepitus redux*.   Dr. Bennett thinks it established that, although absence of the chlorides may be found to exist in some other diseases, and may thus lessen the value of the sign in pneumonia, it yet leaves it unaffected in importance, " as pointing out the onward progress of that disease." (*Op. cit.*, p. 640.)   The appearance of the chloride of sodium—thus excluded from the urine of pneumonic patients—in the *sputa*, and its desertion of the sputa when it again becomes manifest in the urine, is another remarkable fact connected with these observations.   Experiments conducted by Mr. Seymour, Clinical Clerk, upon more than sixty pneumonic patients in Dr. Bennett's wards, established the fact that the chlorides were absent in all but one ; and that was a case of phthisis, with *intercurrent* pneumonia.

The question to which we now recur, is virtually that at the foundation of our present researches, viz., *does any morbid action result from the retention of the chlorides in the blood?*   Irrespective of what we have already said of the avidity with which the chloride of sodium is sought by man and animals, and its innocent nature unless inordinately ingested—when, indeed, it would in all probability be either vomited or discharged from the bowels—we learn from chemical and medical authority that the blood always retains a portion of the chlorides.   The fact of the chlorides being found in large quantity in pneumonic sputa, is explained on this ground by Dr. Thudichum, viz., that sputa being, in part at least, " extravasations and exudations from the blood," the chlorides would naturally appear in a substance partly composed of exuded and " stagnant" blood.   It does not seem, then, from all that we can discover is as yet known, and considered as material from which to draw conclusions, that pneumonia, or the other diseases in which the absence of the chlorides from the urine is remarked as a feature, can be distinctly ascribed to that fact as a prime cause.   Were this to be predicated of any one affection, however, it would certainly be of pneumonia.   An extensive field is open for important and interesting observations in the direction of the present inquiry ; but should it be hereafter ascertained that the amounts of chlorine absent from the urine, and therefore presumed to be circulating in the blood, are to be considered as more or less poisonous, either by their quantity or quality, we still must remember, in estimating such an action, that there are other channels of excretion open, by which a portion of the chlorine is excreted. The constant occurrence of diminution or abolition of the chlorides in pneumonia, is a curious and important fact; but, as yet, not sufficiently illustrated by observations and study, to take the place of an etiological element of disease. Dr. Thudichum states, in respect to the questions under consideration, "the absence of the chlorides in the urine does not necessarily involve the absence of chlorine from exudations. For the latter are products of diseased action derived directly from stagnant blood, and certainly not subject to the specific laws of secretion. The presence of chlorine in sputa, therefore, at a time when it is absent from the urine, is not sufficient proof of a determination of the chloride towards the inflamed lung; a proposition which, moreover, loses all probability from the partial or total disappearance from the urine of the chloride in all acute diseases." (*Op. cit.*, p. 166.) He then enumerates bronchitis, typhus, acute rheumatism, pyæmia, and pleuritis, as examples of this fact; and also refers to the influence exerted in the system by varying quantities of chlorine introduced into it with the food, ascribing no inconsiderable amount of this effect to the sort of diet used by the patient when the quantity of chlorine in "pathological urine" is to be estimated. In most acute diseases, it is known that patients take but little food—often none—and that the articles they eat are often unsalted. Another fact of consequence, which is insisted on by Dr. Thudichum, is, "that urine containing no appreciable trace of chlorine is secreted from blood containing a certain amount of it; from which it follows that the composition of the blood is such as not to allow any further removal of chlorine, or that the kidneys have lost their secretory activity as regards chlorine, as well as (which has been seen to be the case) with reference to water." (p. 167.)

As has been shown to be the fact with regard to pneumonia, so it may, in the opinion of the latter writer and others (and it is no less deducible analogically), be considered available to esteem the amount of chlorine in the urine, a gauge of the amount of morbid action going on in the system in certain other affections. That is, an estimate, more or less accurate, may be made as to the *severity* and *activity* of the disease. The minimum of chlorine in the urine, in making such estimates, Dr. Thudichum places at 0.5 gramme; and after that, in the intensity of the disorder it may, as we have seen, be altogether wanting. "This may be the combined effect of an entire loss of appetite, copious serous diarrhœa, or other serous exudations; of secretions, such as perspirations; and of the want of secreting power of the kidneys. A rise in the amount of chlorine, on the other hand, indicates a steady abatement of the acuteness of the disease, and is a good measure of the returning appetite and improved digestive powers of the patient."

We referred, previously, to the experiments of Barral and others, which seemed to show increased activity of nitrogenous elimination through the

agency of chloride of sodium. Dr. Golding Bird remarks, in this connection, that it would seem that this chloride, "besides furnishing hydrochloric
acid to the stomach, and soda to the bile, also exerts an important physiological influence in aiding the metamorphosis of tissue, and consequent
depuration of the blood." (*Op. cit.*, p. 127.) If we accept this as true,
we not only need not ascribe any morbid effects to the retention of chlorine
—or of chloride of sodium—in the blood, but rather deem its presence
desirable and salutary—at all events, in the vast majority of cases.

Certain of Hegar's conclusions, which were arrived at under the super-
intendence of Liebig and Vogel (G. Bird), are interesting in this connection. We select a few, in illustration :—

"The amount [of chlorine in the urine] varied in different individuals,
depending partly on the food, and partly on habit of life and constitution."
"It was increased by exercise and copious draughts of water, which appeared
to act by washing it out of the system, as the augmentation was only temporary." "Indisposition diminished the quantity." "In health, though no
chlorides were taken with the food, they were always found, and must therefore
have been obtained from the blood or tissues." "When a larger quantity was
taken than usual, the whole did not escape from the system by the kidneys, nor
even the bowels."

The latter two facts would seem to go far to prove the innocuousness of
even considerable amounts of chlorine in the blood. Dr. Day (*Contribu-
tions to Urology*) is quoted by Dr. Bird as testifying to the fact to which
we have previously called attention ; that the chlorides are diminished in
all cases of disease accompanied by copious exudation from the blood.

We notice that Dr. Bird asserts, as from Dr. Beale, that the nature of
the diet necessarily adopted in pneumonia and other acute diseases, seems
insufficient to explain the absence of the chlorides from the urine. Doubtless this default of ingestion of chlorine will not fully meet the requirements of the case in the way of explanation, but it seems only reasonable
to allow it no inconsiderable weight. The following are Dr. Beale's very
important propositions :—

"1. That chloride of sodium is totally absent from the urine of pneumonic
patients at the period of complete hepatization of the lung.
"2. The chloride reappears after the resolution of the inflammation.
"3. The chloride exists in the blood in the largest quantity, when most
abundant in the urine, and *vice versâ*. [We may here, again, find reason for
inferring the harmlessness of even large quantities of chloride of sodium in the
blood ; for when it is considered that under these circumstances, *were it easily,
or at all, a cause of disease*, the conditions for the development of morbid
action abundantly exist, we certainly have sufficient ground for the above
opinion.]
"4. The chloride exists in very large quantity in the sputa of pneumonic
patients.
"5. There is reason to believe that in pneumonia the chloride is determined

towards the inflamed lung, and is re-absorbed and removed on the resolution of the inflammation."

The validity of this latter proposition, as we have previously stated, is denied by Dr. Thudichum; and if it is meant thereby, as it would seem to imply, that the chloride of sodium is the morbific material, we cannot see any sufficient reason to suppose such a relation of cause and effect as by any means certain. That there is a degree of plausibility about it, we confess; but not, at present, any satisfactory proof.

In chronic diseases, the amount of chlorine excreted is usually diminished. This would tend to ratify the supposition that failure of the appetite, and the consequent less ingestion of food, explain the decrease in the chlorine introduced into the system; and the fact consorts, also, with the enfeebled nutrition. Lecanu found the quantity of chloride of sodium very small in the urine of women and old men (*Simon's Chemistry*, vol. ii. p. 167. Sydenham Society's edition.) Observers point out *diabetes insipidus* as an exception to this rule; the chlorine discharged being in excess. So, in dropsy, Vogel found the chlorine increased under artificial diuresis. Chlorine here becomes a gauge of the powers of digestion. It may be inferred that digestion is in good order, when from six to ten grammes are excreted in twenty-four hours; any quantity below five grammes, for the same period, declares an impaired nutrition, unless the diminution can justly be ascribed to a diet very deficient in chlorine, or entirely without any. So those discharges which diminish the amount of chlorine contained in the blood, "as serous diarrhœa, exudations and perspirations," should be taken into account in making an estimate of the morbid action. When the amount of chlorine is very largely increased, and there has been no corresponding plentiful supply by ingestion, *diabetes insipidus* may be inferred. "In dropsical and hydræmic conditions, an increase of the amount of chlorine is a favourable symptom." (Vogel, by Thudichum, *op. cit.*, p. 168.)

If this portion of our subject seems to have been presented in a somewhat negative manner, it is because the information attainable in reference to it is of the same nature. We seem able to say rather wherein the presence of chlorine in the blood—at all events of such amounts as are derivable from diverted excretion thereof by the kidneys—is harmless or even beneficial. Notwithstanding, it must be confessed, that observations and researches upon this point are not available in such quantity as to enable any one, at present, to set forth entirely reliable practical rules and conclusions.

SULPHURIC ACID. (Formula : $SO_3 + HO$[1] : Equivalent, 16.0 Sulphur ; 24.0 Oxygen $= 40.0$, Sulphuric Acid.)

A variable, and often very considerable amount of sulphuric acid is dis-

---

[1] " Being the hydrate of sulphuric acid." (Thudichum, *op. cit.*)

charged from the body by the kidneys during each twenty-four hours. When it is considered that sulphur must nearly always exist in the blood, in greater or less amount—being derived from the food—it will be conceded that it is unlikely that any retention of it therein, even after its oxidation, unless it be present in very large and improbable quantities, would be influential in inducing actual disease. We will examine the known relations of sulphuric acid, as contained in the urine, to the system—so far as facts enable us to present anything worthy of credence and likely to prove of service.

Liebig, referring to Wöhler's experiments, ascribes the sulphuric acid generated in the system, to the action of the oxygen of the atmosphere upon the sulphur introduced into the blood through the medium of food—and this is particularly true of the albuminous portion thereof, which, of course, constitutes, as a general thing, a large amount of the whole. This process of oxidation, and consequent manufacture of sulphuric acid in the system, carried on as it is through the medium of the blood itself, seems to confirm our previously announced opinion, that this acid, in itself, is innocuous. If a very large amount were long and habitually retained in the circulation, it will be easy to understand that mischief might ensue. Before examining any possible or probable pathological issues, we will further allude to certain physiological facts connected with the existence of sulphuric acid in the system, and its excretion from it. It should be said, before proceeding further, that the oxidizing process above mentioned accounts for the surplus of sulphuric acid which is eliminated from the body, over and above what is ingested with the food, in the form of sulphates. (Liebig, Simon, Thudichum.) It is therefore evident that a full animal diet will, by introducing more sulphur into the blood, increase the amount of sulphuric acid in the urine—and that, under the same conditions, more must, for a time at least, be contained in the blood. The experiments of Lehmann on himself show conclusively the effect of different sorts of diet in increasing or diminishing the quantity of sulphuric acid passed in the urine. Thus, while on a "mixed" diet, 7.026 grammes of the acid were collected during twenty-four hours; when animal food was exclusively used, the quantity rose to 10.399 grammes in the same time; when living on vegetables exclusively, only 5.846 grammes were obtainable in the course of one day. Dr. Thudichum, commenting upon these quantities, pronounces them very high, and is inclined to attribute this to Lehmann's appetite. He was, it is true, in fine health and of robust constitution, and therefore some allowance is to be made; but, as Dr. Thudichum also implies, the intrinsic value of the experiments is unaltered by the mere amounts. The *ratio* of formation of the acid, under the differing influences, is the essential point. Certain very conclusive experiments by Vogel, Clare, and others upon this subject, are given in some detail in the work of Dr. Thudichum. The results are the same in character as those obtained by Lehmann.

There exists, as yet, no test or other means of revelation, so far as we are aware, which can inform us what amount of sulphuric acid may be required by the system in its different states. We know the beneficial effects derived from its medicinal use as a tonic, refrigerant and astringent ; and thus it would appear that in certain states of the constitution it is especially suitable—we may, indeed, say indicated. Whether a certain amount of the sulphates must enter into the circulation in order to perfect secretion, " or whether sulphates may be retained and accumulated in the economy," is, at present, unknown. Dr. Thudichum, while stating this fact, mentions that neither rest, nor activity, nor the ingestion of large quantities of water, seemed materially to affect the amounts of sulphuric acid in the urine, in certain of the experiments of Clare and Gruner. Vogel, however, he states, believed "it probable that such influences exist, that the secretory activity for sulphuric acid is dependent upon certain individual and cosmic influences" (*op. cit.*, p. 175) ; and this is rendered nearly certain by the fact of the difference in the rapidity of oxidation of the introduced sulphur observed in various persons. A very significant opinion of Vogel, quoted by Dr. Thudichum, is of importance as affording a valuable suggestion closely connected with our present inquiries, and in respect to which we need more extended observation. It will be seen, as we just remarked, that the suggestion bears directly upon the point which now engages our attention ; and all inquiries of this nature must tend greatly to advance our knowledge of any diseased conditions (and of the remedial measures suited to them) which may reasonably be supposed to depend, even remotely, upon an undue accumulation of the products we are considering, in the blood. The statement to which we have reference is as follows :—

" Vogel, also, from observation, is of opinion that the prolonged use of sulphates in digestive doses is decidedly weakening, and believes it probable that this depressing action may be due to an accumulation of the salts in the system. When to this it is added that sulphate of soda in larger doses is an emetic, and sulphate of potash a poison, the question as to the influence of sulphuric acid and sulphates in the urine becomes one of sufficient importance to fix the attention of future inquirers." (*Loc. cit.*, p. 176.)

From the same sources of authority we learn that the amount of sulphuric acid in the urine is diminished in febrile diseases. This, as is plausibly suggested, may, in great measure, be ascribable to the nature of the food taken, viz., mainly vegetable—less sulphur being thus supplied to the blood ; and it must also be remembered that the *quantity* of food is very greatly diminished under such disordered conditions.

Dr. H. Bence Jones (*Medico-Chirurgical Transactions*, vol. xxxiv.) found the amount of the sulphates increased in *chorea*, and in aggravated cases of *delirium tremens ;* and he likewise notes a similar augmentation, both of the sulphates and phosphates, in cerebral inflammation. He ascribes this occurrence to the rapid disintegration of muscular tissue in the former affections, and to excessive and rapid oxidation of the cerebral substance in the latter.

Vogel found, in three pneumonic patients, the amount of sulphuric acid discharged, exceptionally, " above the normal average." In chronic diseases, while the quantity was variable, it was generally below the standard amount. It was not increased under a diuretic action which largely augmented the discharge of the chlorides, as in cases of dropsy. The ingestion of sulphuric acid and sulphates by patients labouring under chronic disease, alone produced any increase in the amount excreted ; and a hearty meal of animal food had the same effect in patients suffering from diabetes.

While we are obliged to speak doubtfully as to any abnormal effects ascribable to the action of various amounts of sulphuric acid retained in the circulation, we can say—as we have already once stated—that such effects, from any increased amount likely to be thrown into the blood by failure of the renal excretory function devoted to separating this acid, are not such as to be defined, with our present knowledge; nor can we believe any very serious results likely to follow under such circumstances. If *all* the urinary elements were retained, or only such as greatly predominate—and which we have previously considered—then the other accidents, already detailed, would very surely mask any minor morbid action ; and if either urea or uric acid predominated, we should have their peculiar concomitant or resulting phenomena, to the exclusion of any weaker manifestations. While, therefore, nothing positive can be charged to the presence of sulphuric acid in excess (more or less) in the system, we are fortunate in having suggestions from reliable sources, and accurate chemical observations from many competent hands. And all will join Dr. Thudichum in estimating very highly the determination of the quantity of this acid in the urine as being a sort of index of "the amount of disintegration of albuminous matters in the system, in cases where the ingestion of sulphur in any form or combination is very low or altogether suspended." He goes on, also, to remark that possibly a degree of correspondence of the acid with urea, in amount, might be found to exist, "supposing their inclination to pass the kidneys to be equally great. But upon this point there are yet doubts." (p. 177.) If both sulphuric acid and urea be largely excreted, we are to infer that very free oxidation is going on in the system, and is due to the ingestion of animal food in abundance. The opposite condition with respect to sulphuric acid, would go to show a diminution in, or entire deprivation of animal and vegetable food. These states may be either constant or accidental. (Thudichum.) We have already hinted that a sudden, temporary increase in the amount of sulphuric acid excreted, would seem to indicate a reception of sulphur, in some of its forms, into the system, and that too in large quantity.

Dr. Thudichum thinks that new analyses of the blood will have to be made, and without incinerating it. This, he states, "destroys the relative

5

proportions of acids and bases in the salts of the alkalies." We may thus hope not only for more accurate chemical information upon these points, but also that new and more abundant pathological inferences will be drawn from such researches. The fact that the production of sulphuric acid is proved, to all appearance, to be restricted to the blood, and that it is formed through the agency of the oxygen of the air, by means of the respiratory function, is highly important in relation to the present portion of our subject. Its being formed, in great measure, in the blood, is favourable to the idea we have before mentioned, viz., that the vital fluid will be more likely to tolerate its presence, even in large quantities. If this opinion should be deemed merely speculative, we can only say that most of the information now existing upon the subject is of the same nature.

Another interesting question, founded upon the seeming fact that the sulphuric acid is in great part formed in the blood, from the aliments—and also referring to the remark by Liebig, "that the acid nature of the urine of carnivorous animals, as well as that of men, depends upon the nature of the bases partaken of in the aliments, and upon the particular form of their combination"[1]—is, how far the office of the kidneys is to finish the "final oxidation, or that stage of disintegration of albuminous matter in which sulphur, in the form of sulphuric acid, leaves the organic combination, joins a base, and appears in the urine." (Thudichum, op. cit., p. 178.) These, and other chemico-pathological inquiries, as has already been intimated, must be left to future investigators to determine. So far as our present means afford us any ground for the pathological inferences connected with this portion of our subject, we are compelled—after having presented what, for the greater part, is as yet conjectural and undetermined, although seemingly interspersed with the elements of truth, and certainly accompanied by many precise chemical facts—to rest the matter here.

PHOSPHORIC ACID.—(Formula : $PO_5 + 3HO$.)[2]

This acid is the next regular constituent of the urine which we are to consider, pathologically. Previous to entering into any particulars, it may

---

[1] Liebig, quoted in Simon's Chemistry, vol. ii. p. 153. (Lancet, 1844.)

[2] "The common or tribasic phosphoric acid."

Chemical Composition of Phosphoric Acid.

1 equivalent phosphorus	P=31.436
8 equivalents oxygen	=64.000
3 equivalents hydrogen :	= 3.000
1 equivalent of phosphoric acid	=98.436

"The theory which assumes P to be a double atom, and the single atomic weight=15.718, uses P_2 as the symbol for the above equivalent of phosphorus. This is an explanation to the reader, should he find himself embarrassed by the formulæ of different authors."—Thudichum.

be said that its position in relation to pathological states of the system, generally, very closely resembles that of sulphuric acid. We shall therefore treat of it in much the same manner as was adopted in examining the latter substance.

Phosphorus enters the system constantly, and often in very considerable quantities. It is taken with the food, and it has also been long medicinally ordered. Within a short time, indeed, the *phosphates* and *phosphites* of lime and soda have been prescribed with variable results, as *nervines*, and also as being suited to combat the ravages of tuberculous disease of the lungs. In certain cases, we may add, they are reported to have been of service ; and we can recall cases of general prostration, and of what has been sometimes termed "nervous debility," in which we have used them with apparently marked benefit. To Dr. Churchill, now of Paris, belongs, we believe, the credit of suggesting their persevering employment in threatened, or actually existing, pulmonary consumption. The success attained, although flattering in many instances—as we learned, personally, not long since, in the French capital—has not, by any means, justified the expectations at first excited.

This seeming digression from the immediate course of our subject, is not, after all, an element actually foreign to its consideration; for we may thus at least be led to examine what cases will be likely to derive benefit from the ingestion of phosphorus into the system.

The amount of phosphoric acid present, normally, in the urine, is considerably less, according to Becquerel and Rodier, Golding Bird, Johnson, Prout and others, than that of sulphuric acid ; and it is found in combination with lime, ammonia, soda, or magnesia. Dr. Thudichum, we notice, in his estimate in the table from which we have taken our list of urinary constituents, has made the amount greater. In the system generally, the quantities existing at different times, will vary from nearly the same causes as have been assigned for the changes in the quantity of sulphuric acid ; viz., the nature and amount of the food taken, and the ingestion or otherwise, of phosphorus in a medicinal form. The same difference in rapidity of excretion of phosphoric acid, is observed in different persons, as is true with regard to sulphuric acid. And in respect to the activity and amount of excretion, the same variations are to be remarked as have been recorded for chlorine and sulphuric acid. Therefore, renal disease must be considered an important *pathological* influence in respect to the amounts excreted, as personal peculiarities and fortuitous circumstances as to diet, &c., are observed to be, *physiologically.* We learn from Vogel, that copious draughts of water will increase the amount of phosphoric acid excreted by the kidneys—a fact similar to what has been mentioned in respect to the chlorides. As Dr. Thudichum remarks, in commenting upon this point, and as we have previously intimated, "the organism may at one time contain an excess of phosphoric acid, at other times the acid may be deficient."

"It will, however," he adds, "be difficult fully to establish these points, until the normal amount of phosphoric acid contained in all parts of the body, and its changes and variations, within the range of perfect health, be known. And then the examinations will have to comprise a complete analysis of all food, and of all excretions." (*Op. cit.*, p. 188.)

The decrease of phosphoric acid in acute disease, noticed by Vogel, is ascribed partially, as was the fact with regard to sulphuric acid, to the *diminution* and *quality* of the diet ordered or made necessary. When the rations are richer and more liberal, the phosphoric acid increases in quantity; and in convalescence the normal amount is often exceeded, owing to the increased ingestion of food. The decrease of phosphoric acid bears a notable proportion to the *period of time* the illness lasts—even if the attack be violent, and much fever accompany. If short, there is little variation in the amount of acid excreted, and *vice versâ*. Severity and prolongation of the disease, with great diminution of, or total abstinence from food, causes very marked decrease in the quantity of phosphoric acid appearing in the urine. In *chronic diseases*, no rule is observed. Sometimes, and indeed usually, there is great diminution in the amount, and again there may be excess.

Certain statistics supplied by Dr. Thudichum, relatively to the quantities of phosphoric acid excreted in various diseases, and at different stages of each affection, have an important bearing upon the subject. In a table given by him (*op. cit.*, p. 191), the most remarkable points, perhaps, are that the largest amount mentioned ("maximum") as excreted, was in the case of a *female* with diabetes insipidus (7.8 grammes); the next largest, in a *male*, suffering from *hydruria* (5.8 grammes). This seemingly ratifies the remarks and observations previously made, that where the kidneys are most actively employed, the most phosphoric acid is excreted.

If the question be now asked, whether there is any relation between the deficient excretion of the acid in question, and the diseases which have been specified as exhibiting less of the excreted product during their acute course —this implying the retention of a greater or less amount of the substance in the blood—we must admit that no positive proof of this absolute connection as yet exists. In this respect, as in many others pathologically important, the subject is very much in the position of the one last examined. We have already, incidentally, referred to the detection of an increased amount of the phosphates—in connection with the sulphates—in the urine of persons with cerebritis, by Dr. H. Bence Jones. This phenomenon, ascribed by that accomplished observer to rapid oxidation of the cerebral tissues, will, we conclude, hardly justify us in ascribing the inflammatory attack upon the brain, in any degree, to increase and undue retention of the phosphoric acid in the circulation. It does not appear, moreover, that the kidneys were at all disabled; had they been, it is not impossible but more serious mischief might have occurred in such cases—although it is known that a portion of the phosphorus entering the system, goes off by the bowels, as well as from the kidneys.

EARTHY PHOSPHATES.[1]

We have previously mentioned the medicinal use of the phosphates, with the purpose of adding to the nervous force, and in the hope of obviating tuberculous disease. The phosphates of lime and of soda have thus far been prominent in these respects. The latter is physiologically essential to the integrity of the blood and body.[2] (Liebig, Thudichum.) The proportion of earthy phosphates discharged daily is found to be very variable in different persons; and, according to the most reliable authorities, no average amount can yet be declared. Lehmann, when living upon a mixed diet, discharged, on the average, 1.09 grammes, daily; when his food was exclusively of an animal nature, the amount became 3.56.[3]

[1] *Chemical Composition of the Phosphates.*

Ammonio-phosphate of soda$=PO_5+NaO+NH_4O+HO+8HO.$
Acid   "   "   " $=PO_5+NaO+2HO.$

Ammonio-phosphate of magnesia$=PO_5+2MgO+NH_4O+12HO.$

Phosphate of lime (acid),$=PO_5+2CaO+HO.$
Phosphate of magnesia, $=PO_5+2MgO+HO.$

Phosphates of the Alkalies, and Alkaline Earths.

Thudichum.

Phosphate of soda . . . . . $(HO,2NaO,P_2O_5)+24HO.$
Ammonio-phosphate of soda . . . $(HO,NH_4O,NaO,P_2O_5)+8HO.$
Phosphate of lime . . . . $(HO,2CaO,P_2O_5)$
Ammonio-phosphate of magnesia . . $(NH_4O,2MgO,P_2O_5)+12HO.$
Neutral phosphate of soda . . . . $(HO,2NaO,P_2O_5)+26HO.$
Acid phosphate of soda . . . . $(12HO,NaO,P_2O_5)+2HO.)$

G. Bird.

Robin and Verdeil have pronounced the two salts, last described, to be normal constituents of the urine.

[2] "There is no known salt the chemical characters of which approach more closely to those of the serum of blood than the phosphate of soda; there is none more fitted for the absorption and entire removal from the organism of carbonic acid."—Liebig, Researches on the Chemistry of Food, and the Motion of the Juices in the Animal Body. Dr. Gregory's Translation, American edition, by Prof. E. N. Horsford, 1848.

[3] In connection with the presence of phosphoric acid and of chloride of sodium in the system, the following remarks of Liebig have an interest and importance: "In some pathological conditions there has been observed (Schmidt, *Annalen der Chimie und Pharmacie*, vol. lxi. p. 329), at points where bones and muscles meet, an accumulation of free lactic and phosphoric acid, which has never been perceived at those points in the normal state. The solution and removal of the phosphate of

Pathologically, the effect of the presence of an unusual amount of the earthy phosphates in the system, would depend very much upon the integrity of the kidneys. If the latter were disabled, and even if not particularly disorganized—sometimes, even, if nearly healthy—there might be large deposition of these substances in various parts of the urinary passages. The circulation would in this manner be freed from an embarrassing amount of them, but the morbid effects of such an abundant deposit would certainly prove very troublesome. Vesical calculi are rarely *entirely* composed of the phosphates. Such is their friable nature and tendency to a pulverulent condition, that they prefer a nucleus of some sort, around which to accrete, rather than to form pure aggregations of their own substance.

Dr. Thudichum believes that "the originators of the term phosphatic diathesis and phosphuria, and their followers, linked a series of the most varied disorders together under this term, which had nothing in common but one symptom, namely, alkaline urine." (*Op. cit.*, p. 211.) He then proceeds to explain in a very clear manner, the reasons for a greater or less acidity or alkalinity of the urine. And alkalinity may thus be often due to a lack of animal food, by which a suitable amount of acid is usually supplied. Therefore the invalid and the poor man are alike liable to pass very alkaline urine—the dyspeptic from want of appetite, and the pauper from want of means. A vegetable or fruit diet, alone, will make the urine alkaline; and those who cannot digest, or imagine they cannot digest meat, and therefore try a vegetable diet, will almost certainly have alkaline urine. In anæmia, a meat-diet will soon produce an acid urine, with the phosphates, where no acidity existed before.

It would appear, then, so far as conclusions can at present be drawn, that a certain amount of phosphoric acid is not only physiologically necessary (*phosphate of soda*), but that certain states of the system require additional amounts medicinally, or by food. The ammoniacal urine of patients who labour under spinal disease or injury, or who are suffering from other affections giving rise to retention and stagnation of the urine, and its alkaline change, is referable to a local cause; and the mischief which undoubtedly results to the circulation by reason of this state of things, is

lime, and therefore the disappearance of the bones, is a consequence of this state. It is not improbable that the cause, or one of the causes of this separation of acid from the substance of the muscles is this—that the vessels which contain the fluid of the muscles have undergone a change whereby they lose the property of retaining within them the acid fluid they contain.

"The constant occurrence of chloride of sodium and phosphate of soda in the blood, and that of phosphate of potash and chloride of potassium in the juice of flesh, justify the assumption that both facts are altogether indispensable for the processes carried on in the blood and in the fluid of the muscles." (*Op. sup. cit.*, p. 90.)

part and parcel of the causative disease. If inordinate quantities of the phosphates are passed for a long time, we should certainly look for systemic disorder ; and doubtless the general health would soon be found to deteriorate. Some *specific* cause might be detected for such failure—unless an unusual quantity of phosphoric acid were being introduced into the system, to account both for the large corresponding excretion, and for unimpaired health—should the latter be observed to exist. But usually, with such a state of things, impaired nervous energy, dyspepsia with irritability, and various functional disturbances, will be present.

The extraordinary case reported by Dr. Golding Bird, of the man who passed very large quantities of the phosphate of lime, without apparent harm to his constitution (excepting that he was always dyspeptic), is believed by Dr. Thudichum to be an instance of imposture. There certainly seems to be some reason for so regarding it. The patient had been under the care "of half the hospital physicians and surgeons in London," during fifty years. It is remarked that very possibly he might wish to be an object of permanent interest to whatever physician attended him, and that he liked hospital quarters. At one time he brought more than an ounce of the above-mentioned salt to Dr. Bird, and which he asserted was passed from himself. His urine was milky, and abundantly deposited the salt, as stated. It is not, however, impossible that he may have been even a longer time collecting the large quantity shown to Dr. Bird, than is suggested by Dr. Thudichum, viz., sixteen days. The man's health was so good that no treatment seemed justifiable, except on account of an apprehension that a vesical calculus might form.[1] Dr. Thudichum does not think Dr. Bird's explanation of the case founded on fact ;[2] and he refers to Dr. Prout's remarks upon this point, as follows:—

"If the reader should not share our doubts, he may adopt the explanation by Dr. G. Bird, for which, however, there is no basis in fact ; or he may explain it upon the ground of the following observation recorded by Dr. Prout. (p. 323, note.) This physician examined the body of a gentleman who, during the greater part of his life, had suffered from renal disease, remarkable for being

---

[1] It would therefore seem that his *dyspeptic* symptoms were not very urgent.

[2] The following are Dr. Bird's explanatory remarks: "In cases of this kind it is very possible that the phosphate of lime is secreted from the mucous membrane of the bladder, and not derived from the urine. All mucous secretions contain phosphoric acid, combined with earthy bases ; and hence if an excess of the latter is secreted with the vesical mucus, it may be washed away with the urine and form a deposit. This is by no means unfrequent in the irritable bladder, depending on the existence of prostatic diseases, &c.: we have a perfect analogy to this in the calculous concretions found in the ducts of glands furnishing mucous secretions. These are all prone to secrete phosphates in too great an excess to be washed away with the secretion ; they are therefore retained and form a calculus. These, from whatever part of the body they are obtained, present nearly the same composition." (*Urinary Deposits.* Dr. Birkett's edition, 1857, p. 306.)

attended by the secretion of large quantities of the earthy phosphates. Both kidneys were not only extensively disorganized, but most of the natural cavities, as well as many cysts, were found distended with numerous earthy concretions, of various sizes and composition. The concretions found in those cavities *to which the urine had access*, consisted of the phosphate and carbonate of lime, and more or less of the triple phosphate of ammonia and magnesia, while those cavities or cysts distinct from the renal structure, and to which, therefore, *the urine had no access*, consisted of the calcareous phosphate and carbonate only, without any admixture of the triple phosphate." (*Op. cit.*, p. 213.)

When what is termed the "phosphatic diathesis" exists, or when we find copious phosphatic deposits in the urine, we are to expect in such patients a state of debility, listlessness and exhaustion, mental and physical —a sort of cachexia, with disturbed digestion, and an irritable state of the digestive organs; and also, either some manifest or concealed disorder of the nervous system. To this condition we have already referred. Such symptoms should lead to the adoption of an alterative and tonic treatment, and to such examination of the patient hygienically and constitutionally, as will doubtless soon afford a knowledge of the chief source of difficulty —whether functional or systemic; and if the latter, whether the cerebrospinal, or renal organs are at fault; or whether the blood itself be surcharged with matters fit only for elimination. Old people often exhibit a train of symptoms indicating troubles referable to what is styled "the phosphatic diathesis." Dr. Bird well describes the state when he writes that there "is irritability with depression, a kind of erethism of the nervous system, if the expression be permitted, like that observed after considerable losses of blood."

It seems peculiarly appropriate to introduce in this place the conclusions of Dr. H. Bence Jones in reference to the pathological bearing of phosphatic salts in the economy—much of what we have previously said seems to find ratification in these opinions, and perhaps we could not better conclude our examination of this department of our subject than by presenting them.

"1. No determination of an excessive secretion of phosphoric acid can be afforded by the deposit of earthy salts, unless the quantity of lime and magnesia in the food be taken into account.

"2. No *real increase* of phosphatic salts occurs in spinal diseases, notwithstanding the existence of deposits.

"3. In fever, and in most acute inflammations, the phosphatic salts are not increased.   .

"4. In old cases of mania, melancholy, paralysis of the insane, or in chronic cases of disease in which nervous tissues are uninfluenced, no conclusious can be drawn.

"5. In fractures of the skull the phosphatic salts increase only when any inflammatory action occurs in the brain, and in acute phrenitis an excessive increase takes place.

" 6. In delirium tremens there is a marked deficiency of phosphates unless they are introduced with the ingesta; an excess is, however, met with in some functional affections of the brain."[1]

In this connection, we cannot but allude to the zealous, and, as it would seem, very sensible, recommendation, by Dr. Bird, of what he appropriately terms "renal depurants." Knowing, as we do, from unmistakable symptoms, that a poisonous substance is traversing the blood, and thus pervading all the bodily tissues, it is our manifest duty to use some such means for stimulating a sluggish, or even a partially diseased kidney, to freer elimination. And there are many substances which may be effectually employed in this manner, and innocuously, or even beneficially to the kidneys themselves. Indeed, when there is even some degree of risk in demanding extra work from those organs, it is better to tax them rather severely, than to allow the accumulation of morbid matters, of any kind, in the blood. "The alkalies, their carbonates, and their salts, with such acids as in the animal economy are capable of being converted into carbonic acid, including the acetates, tartrates, citrates of soda and potass," are properly "renal depurants." (Bird, *Op. cit.*, p. 452.)

AMMONIA.—Symbol : NH_3; Equivalent, 17.0.[2]
Ammonia exists in comparatively very small proportions in fresh and healthy urine.[3] Liebig, indeed, doubts whether it can indubitably be pronounced an invariable and constant constituent of normal urine. In these doubts he is joined by Lehmann and Scherer. Dr. Thudichum, who has investigated the subject in his treatise already largely referred to, thinks the researches of Boussingault more satisfactory than those of Bœcker and De Vry ; and that they tend to show the presence of this substance in the urine. Neubauer (*Journal für Practische Chemie*, Bd. 64, p. 177, and *Anleitung*, § 56), according to the same writer, has afforded the best proof, thus far, of the presence of certain amounts of ammonia in the

---

[1] We transcribe these statements as given in Dr. Bird's volume ; consulting at the same time Dr. Jones's papers on the Sulphates and Phosphates, contained in the 30th and 34th volumes of the Medico-Chirurgical Society's Transactions.

[2] *Chemical Composition of Ammonia.*

$H_3 =$ 3.0	17.65
$N =$ 14.0	82.35
17.0	100.00

Thudichum.

[3] Ammonia "is the only volatile alkali with which chemistry is acquainted ; and of this property we avail ourselves for its analysis."
"*Demonstration of the presence of Ammonia in Urine.*—The ammonia, which has been liberated from urine by means of milk of lime, is made to pass in the form of gas into a solution of sulphate of silver and arsenious acid ; the precipitate ensuing is evidence of its presence." (*Thudichum.*)

urine. His analyses were made by the method of Schlösing. In regard to this process, Dr. Thudichum remarks: "There is only one objection to this method, which I have already advanced; the ammonia is set free by the addition to the urine of milk of lime. Now, if it can be proved that milk of lime, at the ordinary temperature of the air does not within a reasonable limit of time create ammonia from urea and the other organic substances, we are bound to say that an essential progress would be effected by these researches of Neubauer. The subject of ammonia in connection with the animal economy would be of immense importance, if it should be proved beyond the shadow of a doubt, what Dr. Richardson (Astley Cooper, Prize Essay for 1856, On the Cause of the Coagulation of the Blood) has endeavoured to show, namely, that ammonia is a regular constituent of the blood, and the solvent of fibrin in the living body." (pp. 219, 220.)

According to the same author—who again refers to Neubauer's analyses for the proof of the assertion—" *The ammonia of the salts of ammonia, when the latter are taken into the stomach, passes unchanged through the system and is discharged in the urine.*" (p. 223.)

We have cited the above facts—principally of a chemical and physiological nature—because they are not only interesting, but also have an important relation to the pathological question which presents itself in regard to ammonia considered as a constituent of the urine; and whether its retention in the blood produces any morbid effects. We have not very much to offer upon this portion of the subject; but there are some considerations which are significant. There is a great deal to be learned in regard to the matter—to say nothing of the points in dispute—and principally in reference to the production of urœmia through the agency of *carbonate of ammonia*, as a product of fermentation in the blood. Upon this latter question we have already entered into extensive detail, and have endeavoured to present the actual *status* of the subject, as viewed by many celebrated and industrious observers. We refer the reader to the portion of this essay devoted to *Urœmia;* and will merely add, that all who have examined the facts and theories advanced by Frerichs and others relative to urœmic poisoning, acknowledge the importance of the investigations, and also seem fully aware of the conspicuous position which ammonia would assume as a morbid agent in the human system, if it should ever be proved to possess such a toxic influence as some now accord to it. Dr. Thudichum remarks upon this point: "If what some have ventured to bring forward as a defined feature of certain forms of disease of the kidney can really be maintained, namely, that the urea retained in the blood may there undergo decomposition into carbonate of ammonia, and give rise to the symptoms described as urœmia, the pathological indications of ammonia in the urine would be all-important in those diseases. And though quantities of ammonia might be excreted by the lungs, skin, and bowels, yet the urine would be that excretion in which the ammonia would be most accessible. However probable such a process, under given circumstances, may be, actual and direct proof would be required to make it a fact; and this we cannot say to have been afforded by the originators of the theory. We know, on the contrary, that the test said to be diagnostic of the presence of ammonia in the breath, the formation of white vapours on contact of the breath with a glass rod dipped in hydrochloric acid, frequently fails in cases with the most marked symptoms of uræmia. We must, therefore, expect further proofs, analyses of the blood and the excretions, before we can give that extension to toxæmia as a cause of various severe affections, which by various authors has been attributed to it. It is the same with putrid or septic fevers, under those conditions in which the blood is said to be in a state of dissolution. For all we know, ammonia may be a product of these pathological processes; and then we might expect to find it, in part at least, in the urine." (*Op. cit.*, pp. 224, 225.)

The opinion of M. Claude Bernard, in his late work (*Leçons sur les Propriétés Physiologiques et les Alterations Pathologiques des Liquides de l'Organisme*, Paris, Baillière, 1859), is adverse to admitting that carbonate of ammonia is capable of producing toxæmia. We have referred to this opinion more at length in another place. (See Appendix, Note A.)

The small quantity of ammonia entering into the composition of the normal urine, is, of itself, an element which rather tends to preclude the idea of its accumulating in the blood—in cases of retention of the urinary ingredients—in such quantities as to prove noxious, even if we consider it to be a toxic agent. At all events, under such a supposition, it would require a considerable time for bad effects to arise. And even granting all these conditions, the phenomena springing from retention of the other more abundant constituents would preponderate; unless, by that sort of "elective affinity" previously spoken of, ammonia alone were seized upon and retained in the circulation—the other matters being eliminated and excreted.

Again, it is by no means impossible, although not definitely proved, that ammonia may be essential to the integrity of the blood; in which case, any pathological inferences from its presence, unless occurring in very large and improbable quantities, could hardly be drawn. And especially in view of the fact that the nature of that ferment supposed by Frerichs to be necessary for the production of carbonate of ammonia from urea contained in the blood, is yet unknown, must doubt envelop the whole question, pathologically, until further experiments shall be made, and its existence and real character be ascertained, or its nonentity determined.

In respect to the question as to the necessity of ammonia to the proper constitution and healthy condition of the blood, Dr. Thudichum has some remarks which, under cover of a little pleasant facetiousness, contain valuable hints, and may be appropriately introduced in this connection. Speaking of the elimination of ammonia—ingested into the system in an unchanged state—by the kidneys, he says :—

"It remains to be seen whether caustic ammonia and carbonate of ammonia are eliminated in a similar manner. It remains, also, to be ascertained whether the organism produces any ammonia under ordinary circumstances, or whether the ammonia in the urine is simply introduced by our food and drink, or by the air which we breathe. Some articles of food are rich in ammonia, e. g., radishes. The smoke of tobacco contains a large share of ammonia; and any person remaining for any length of time in a room filled with this ambrosial offering to Apollo, must inhale such quantities of ammonia as must materially increase the ordinary amount in his urine. If ammonia be really essential to the blood, the anti-tobacco leaguers may yet hear the argument advanced, that tobacco-smoking is really essential to keep our fibrine in solution, and that smoking has of late become so much more common because the ordinary sources of this 'food,' the cesspools, dunghills, and other like accompaniments of human and animal habitations, have been done away with. A still greater amount of ammonia is of necessity inhaled where both the sources just mentioned flow without restraint." (Op. cit., p. 224.)

The fibrine in the blood of a very considerable proportion of the population of America ought to be in a good state of "solution" if the smoking of tobacco be in any degree conducive thereto ! Let us hope that such is the effect of "the weed," viâ combustion and inhalation !

Several observers have stated the very small amount of ammonia which can be ascertained to exist in healthy urine.[1] Simon (Chemistry, Syd. Soc. edit., vol. ii. p. 132) says it "cannot be very easily demonstrated in healthy urine." Liebig (Lancet, 1844), referred to by Dr. Day, Editor of Simon's Chemistry (loc. cit.), pronounced the presence of ready formed ammonia in the urine, as only indicated by "very minute or doubtful traces;" and stated also that "these traces probably pre-existed in the food partaken of." Dr. Day subsequently remarks : "Experiments for the determination of the amount of ammonia in the urine of healthy individuals may become of importance in judging of pathological states ; for in fevers and other diseases, the amount of ammonia in the urine increases considerably. It is possible that by analyzing the urine we may, in the increasing or decreasing amount of ammonia, obtain a measure for the alterations which take place in diseases."

As has been previously remarked, there seems not to be a sufficient amount of evidence from which to educe satisfactory conclusions relative to any pathological influence ammonia may have upon the system when present therein in unusual quantity. Those who attach such a distinct power to it in the production of uræmic intoxication and eclampsia, have yet to

---

[1] "Ammonia exists only in the urine in combination with the muriatic, phosphoric, and lithic acids." (Prout, op. cit., p. 555.)

procure and offer much additional evidence, before their doctrines can be unhesitatingly and fully accepted. The preliminary structure certainly has a fair and plausible appearance, but needs full development and confirmation. We have endeavoured to present whatever is available and reliable upon the present topic, so far as our means of information permit.

The examination of the relations of ammonia to the urine and the blood concludes the plan originally proposed by us in discussing the subject under consideration. It occurs to us, however, that a few words may appropriately be added relative to the *iron* and the *colouring matter* of the urine— *uræmatine.* These we will briefly notice under one head.

IRON AND URÆMATINE.[1]

Iron and uræmatine exist normally together in combination[2] in the urine. Of course, variable amounts of iron must be found at different times in the system, according to the nature and amount of the ingesta ; and as the mineral in some of its various medicinal forms is very extensively used, there must be in many persons a large quantity introduced into the circulation.[3] We know the marked beneficial effect of this medication in a large class of cases, especially in those of an anæmic and chlorotic character, and its employment was doubtless owing, first and mainly, to the knowledge of the fact that iron is a normal constituent of the blood. That portion of the vital fluid of which it is an integral part, is, as is well known, the *hæmatine* or red colouring matter. Chemists have been inclined to believe uræmatine "a derivate" of hæmatine, so great is its resemblance to it. (Thudichum *et al.*)

But little that is positive can be said upon this subject, as to any pathological deductions. We can hardly conceive of a sufficient quantity of iron being accumulated and retained in the blood—*if derived merely from the amounts contained in the urine*—to prove injurious. It is not infrequently observed in practice, that only certain quantities of iron can be medicinally ingested, without giving rise to headache, fulness and turgidity of the bloodvessels, with other disagreeable symptoms, principally of a plethoric

---

[1] *Iron:* Symbol ; Fe. Equivalent ; 28.0.

Uræmatine: Elementary composition unknown.

Thudichum.

[2] Harley ; cited by Thudichum.

[3] Notwithstanding that iron is frequently thus largely ingested, it is a curious and hitherto unexplained fact, that it is not found in the liquid excreta. Its mode of elimination is therefore unknown. Upon this point, M. Cl. Bernard remarks: "Iron is the only substance which exists normally in the blood, which has not been found in the excreted liquids. How is it eliminated ? This is undecided. A small quantity only is absorbed even in medicamentary usage. It has been found in the hair ; but this is a very slow method of elimination." (*Leçons sur les Propriétés Physiologiques et les Alterations Pathologiques des Liquides de l'Organisme,* Paris, Baillière, 1859, vol. i. pp. 448-9.)

nature, and which necessitate, of course, a suspension of the medicine, or, at least, a marked diminution of the amount administered. Certain persons, also, are far more easily affected by chalybeates than others. In some, this amounts to a species of idiosyncrasy which precludes the use of such remedies, unless combined with other articles whose employment is often not desirable, and sometimes is inadmissible, for peculiar and varying reasons. Thus, we know persons whom iron, given in any considerable quantity, will *purge* persistently; and in some such instances, if opium be combined with it for the purpose of restraining the action of the bowels, *nausea* takes the place of the diarrhœa—a species of intolerance of the medicament seems to exist. This is doubtless not uncommon. There are other manifestations sufficiently indicative of the powerful action of iron upon the system. But such phenomena, and those more especially mentioned above, are distinctly referable to considerable quantities of iron thrown into the organism at once, or gradually. As we have intimated, there appear to be no sufficient *data* whereupon to found any conclusions in reference to a possible pathological result of retention in the blood *of the iron derived from the urine.*[1]

With regard to the curative properties of iron in certain affections, the statements of Dr. Golding Bird have a peculiar fitness to our subject. He says:—

"Among the remedies which appear most successful when food is not converted into healthy chyle, and an unhealthy state of the blood from the presence of imperfectly assimilated matters results, the preparations of iron deserve notice. I have repeatedly seen copious deposits of uric acid, in persons of low power, completely disappear *pari passu* with the cure of the pseudo-chlorotic symptoms present, by the use of this important drug."

This is literally "killing two birds with one stone," enriching the blood, and depurating it, also. There can be little to fear from *iron* when judiciously administered; it is to be hoped that any "morbid effects" from large amounts existing in the circulation, in whatever way this may be produced, will by and by be satisfactorily and fully ascertained and explained. It would seem, however, that, had any very marked pathological phenomena been constantly, or even frequently, specially dependent upon accumulation from urinary obstruction or failure of due elimination, the acumen of the numerous competent observers of disease, so continually on the watch for such manifestations, would, ere this, have given us some positive ideas upon the subject.

Uræmatine.—There are some interesting observations relative to *uræmatine*, which it seems desirable to present at this time. Uræmatine is believed to be derived, in great measure, from the hæmatine set free during

---

[1] "The analysis of the urine may prove useful for determining the amount of iron which enters the blood and circulation when it is taken as a medicine." (Thudichum.)

the disintegration of the blood-corpuscles, which is always taking place. *Cholæmatine*, the colouring matter of the bile, is also considered a product of hæmatine. The colour of hæmatine is known to be very persistent. The colour of the fæces and urine is ascribed by Dr. Thudichum to "effete hæmatine." He also alludes to the fact that iron is always conjoined with uræmatine as a component element, and considers this as confirmatory of its origin from the colouring matter of the blood-corpuscles.[1]

Vogel has shown that the amount of uræmatine is increased in acute febrile diseases; in those which partake of an anæmic or chlorotic nature there is less colouring matter discharged. This is ascribed to the diminution of the disintegration process constantly going on, as has been said, with greater or less activity. In typhoidal conditions, and in fevers of a septical type, there being great dissolution of the blood, the colour of the urine is heightened.

When hæmato-globuline appears in the urine, it is indicative of destruction of the blood-corpuscles, and is a sign of serious import. If only transitory, however, and not recurrent, the case becomes more hopeful. "But when it is a symptom of severe scorbutic or septic disorders, it is a sign of great danger to the life of the patient. Suppression of the urine, and discoloration of the skin, when following the discharge of urine rich in hæmatoglobuline, are also very unfavourable, and are forerunners of a fatal termination of the case." (Thudichum, *op. cit.*, p. 237.)

In discussing a question like that proposed for the subject of this essay, there is much to trammel us, on the score of incomplete revelations of the chemical and pathological elements concerned. The great interest and importance of the theme itself, and of the studies and investigations rendered necessary for its thorough and progressive examination, will more and more rapidly bring hidden truths to light. Practical experiments, and the continued observation of those engaged in medical pursuits, were never more needed than in elaborating the processes and extending the records already so well begun—instituting new methods and employing more profound scrutiny. And especially to those who have the great advantage of being connected with large hospitals, are rare opportunities afforded for these pursuits, destined, as they certainly are, to benefit suffering humanity, by enabling the members of our profession with more facility and certainty, in this class of cases, *prius cognoscere, dein sanare.*

---

[1] For methods of ascertaining the quantity of uræmatine in healthy urine, the amount passed in twenty-four hours, and much valuable information relative to the various shades of colour observed in urine, together with illustrations, see Dr. Thudichum's work. Consult, also, Vogel, "Archiv. des Vereins für gemeinschaftliche Arbeiten," Bd. i. p. 137, 1853.

APPENDIX.

NOTE A.

Urea ; questions relative to its agency in the system.

In a lately published work,[1] M. Claude Bernard discusses the question whether urea is to be considered a toxæmic agent, either *per se* or by any product derived from it. The respect which the opinions of this distinguished observer must always command, and the very recent date of the conclusions at which he has arrived upon this subject, induce us to present such of them as are most essential to our purpose.

"Is urea a poison?" asks M. Bernard, referring to the views of Frerichs. A poisonous substance, he goes on to say, although it may, like urea, be present in the circulation, is not necessarily toxic in every, even the smallest amount. "A violent poison may, then, exist in the blood, in very noticeable proportion, without occasioning symptoms of poisoning, if elimination be sufficiently rapid ; we are not, consequently, authorized to declare that urea is not a poison because it is formed in the blood in considerable quantities." (Vol. ii. p. 33.) Bernard then refers to the experiments of M. Gallois, who found the injection of urea into the blood, innocuous, even in large quantities. We must, then, he continues, admitting the fact of innocuousness, seek another explanation. The theory of Frerichs does not seem sufficient, according to Bernard, to account for the phenomena observed. If carbonate of ammonia be injected in small quantity into the vessels, there is no result ; and even when introduced into the veins of animals in larger amounts, although an extreme agitation was produced, yet life was maintained. Carbonate of ammonia, moreover, has nearly always been found both in pathological and in healthy blood. "Consequently, its presence in the blood cannot explain the special accidents of uræmia." Bernard therefore adopts another explanation, and puts forward the following views in preference to the others referred to. 1. The condition of the kidneys in advanced renal disease ; there is softening, breaking down of the tissue— "fonte putride." 2. By destroying the nerves going to the kidneys, their disorganization is produced ; softening and purulent formations are brought on ; pus is thrown into the circulation ; and thus renal nutrition is wholly perverted, the kidneys becoming decomposed. In this state of things, Bernard is inclined to ascribe the poisoning of the blood and the resulting phenomena to the agency of the putrid matter thrown into the circulation from the injured kidney. He admits, however, that new experiments are necessary. Thus, we ought to see what effects will be produced by injecting the putrid matter from a kidney

---

[1] "Cours de Médecine du College de France. Leçons sur les Propriétés Physiologiques et les Alterations Pathologiques des Liquides de l'Organisme." Par M. Claude Bernard, Membre de l'Institut de France, Professeur de Médecine au College de France, Professeur de Physiologie Générale à la Faculté des Sciences, etc. etc. Paris, J. B. Baillière et Fils, 1859.

which has become disorganized by artificial destruction of its nerves, into the blood of a healthy animal—in fact we must ascertain whether the accidents supervening under such conditions will put on the aspect of the nervous phenomena of uræmia.

"There is no physiological office (*rôle*) known as especially appertaining to urea ; it is a purely excrementitious substance, not a secretion. It is regularly eliminated by the kidneys; when this elimination is interfered with, we observe the supervention of grave phenomena, without being able to say whether they are, either primarily or secondarily, the consequence of an accumulation of urea in the blood, or whether they are dependent upon the lesion which has caused the accumulation." The phenomena referred to, says Bernard, have received the name *uræmia*. "Is it proper to class under the same category the convulsions of lying-in women?" [as we have seen has been done by Braun and others.] This is a question, according to Bernard, which is yet to be examined, and which it would be premature to pretend positively to answer at present. *En resumé*, Bernard thinks the accumulation of urea, or of carbonate of ammonia, in the blood, cannot explain the disorders alluded to ; but that it is far more plausible to believe them referable to destruction or injury of the renal nerves.[1] [This, however, would not explain the so-called uræmic convulsions of pregnancy, since a sort of artificial, or curable, Bright's disease—curable by the act of parturition—may then exist. The question is yet in dispute.]

Dr. E. Brown-Séquard, in the number of the *Journal de la Physiologie de l'Homme et des Animaux*, for January, 1859, makes the following comments upon Dr. Hammond's experiments and conclusions ; and we are happy to avail ourselves of this distinguished physiologist's opinions upon this subject, especially in view of the very recent date of their announcement. Dr. Séquard, after giving Dr. Hammond's views and processes, as published in *The North American Medico-Chirurgical Review* for March, 1858, goes on to say: "We shall confine ourselves to the following remarks : 1. The author has not shown so conclusively, as he seems to suppose, the non-conversion of urea into ammonia (he employed hydrochloric acid as a reagent); 2. Injection into the jugular vein, of substances whose action is sought for, is a procedure very liable to make us suppose a non-poisonous substance poisonous, by reason of the disturbance which the injection of any liquid whatever into this vein sometimes produces in the movements of the heart ; 3. In dogs from whom the kidneys have been removed, the injection of four ounces of pure water is capable of producing the same morbid effects which the author observed after injecting the different salts and urea, contained in four ounces of water. We conclude, then, that Dr. Hammond's experiments, exceedingly interesting though they are, are not sufficient to overthrow the ingenious theory of M. Frerichs. Of the two divisions of this theory, that which considers urea not to be a poison, or, at least, not to be capable of producing uræmia, seems scarcely to be shaken by the experiments of Dr. Hammond. As to the other division, according to which uræmia is a poisoning due to the carbonate of ammonia, more powerful arguments than those we have cited above have already been urged against it, and, at the present

---

[1] For Bernard's remarks, *in extenso*, see the work cited, vol. ii. pages 34—37 inclusive.

time, it seems very probable that uræmia is an aggregation of symptoms dependent upon various causes, among which, poisoning by carbonate of ammonia has only a variable proportion of influence.

"The experiments of M. Gallois (*Thèse inaugurale*, Paris, 1859) may be considered more positive than those of the American physiologist, in showing that urea is a poison. M. Gallois has seen rabbits die, after violent convulsions, from the administration of twenty grammes of urea, introduced into the stomach. But we have no proof that urea is not transformed into carbonate of ammonia, or into some other toxic element." (*Loc. cit.*)

M. Guérard (*Dict. de Médecine*, in 30 vols., Art. *Sang*), referring to the accumulation of urea in the blood, thus writes: "This accumulation in the blood of a principle whose elements are endowed with such an excessive mobility, and which separate themselves with the greatest facility in order to transform themselves into carbonate of ammonia, eminently deserves to arrest the attention of physiologists, on account of the accidents to which it will unfailingly give rise." He then asks, "if it be not probable that the ammoniacal odour pervading the excretions of those who have long had suppression of urine, is the result of the metamorphosis mentioned? The plausibility of this opinion becomes greater if it be remembered that many observers have assured themselves of the existence of urea in the fluids effused into the various serous cavities, in persons who have died of Bright's disease." M. Guérard refers, in this connection, to Dr. Babington. (*Cyclopædia of Anatomy and Physiology*, Art. *Blood*.)

Note B.

The Relation of Uric Acid to Gout.

Dr. Barlow (*A Manual of the Practice of Medicine*, London, 1856), referring to the explanation given by Dr. W. Budd (*Medico-Chirurgical Transactions*), in respect to the elective affinity manifested by certain morbific matters for certain organs and parts of the body, speaks of the support which the humoral pathology of the various gouty affections may reasonably derive from these views; but he is by no means inclined to allow uric acid the important position of chief agent in the production of gout. He says: "It is not, however, intended to imply by what has been said, that lithic or uric acid is the alone or efficient cause of gout; for if this were true, we should always have gout when uric acid is in abundance [not so, we would remark, if it be in process of elimination, and is not *retained* and *accumulated in the blood*—for such are the elements evidently required for the production of the disease], and never have gout without it; whereas, the presence of the one without the other, especially of uric acid without gout, is a matter of everyday experience." [For the reason above alleged; for if uric acid appear in abundance in the urine, we infer from the fact alone, as a rule, its free elimination; consequently, it does not accumulate in the blood, and gout does not occur.] Dr. Barlow continues: "Upon what the gouty diathesis or susceptibility depends we know nothing [?]; it manifests itself in the system by an affinity for the gouty poison (uric acid it may be) in different parts. If this diathesis be such as to produce an affinity of extraordinary intensity, there may be a local excess of this substance, without any such excess, or even with a deficiency, in the system at large, just as there

may be a local hyperæmia, although the general state of the system may be
anæmic. This explanation of the difficulty is merely suggested as possible, not
enunciated as certain; it nevertheless derives confirmation from the recent observation of Dr. Garrod that uric acid is present in the serum, effused when a
blister has been applied over a joint affected with gouty inflammation." (*Op.
cit.*, English edition, 1856, p. 143.) Aside from the force of the evidence adduced by Dr. Garrod in the test referred to by Dr. Barlow, we cannot think the
position of the latter author is strengthened by the analogy which he attempts
to establish—or the similarity which he would imply, in vital action—between
local hyperæmia in general anæmic states and the deposition of urate of soda
by elective affinity. In fine, in the one case, the blood, although deteriorated,
does not, in the circumstances proposed, contain a poison; whereas, in the latter
case, it does, or the local manifestations thereof would not occur. Mere local
hyperæmia, occurring as above mentioned, cannot be properly compared with
the local exhibition, by vicarious action, of the product of an abnormal amount
of uric acid in the blood.

www.ingramcontent.com/pod-product-compliance
Lightning Source LLC
Chambersburg PA
CBHW021957190326
41519CB00009B/1299